Computational Welding Mechanics for Engineering Application

Computational Welding Mechanics for Engineering Application: Buckling Distortion of Thin Plate and Residual Stress of Thick Plate deals with two special issues in the field of computational welding mechanics: buckling distortion of thin plate and residual stress of thick plate.

Through experiment, theory, and computational analysis, the authors systematically introduce the latest progress and achievements in computational welding mechanics, such as weld buckling in lightweight fabrication and residual stress in HTSS thick plate welding. In addition, they explore its application to address real-world engineering problems in advanced manufacturing, such as precision manufacturing and mechanical performance evaluation.

The book will be of interest to scholars and engineers of computational welding mechanics who wish to represent the welding mechanic response, predict the distribution and magnitude of mechanical variables, or optimize the welding technique to improve manufacturing quality.

Jiangchao Wang obtained his PhD from Osaka University, Japan in 2013 in ship and ocean engineering, with research focused on prediction and mitigation of welding buckling. In 2015, he completed his postdoctoral fellowship in effective computational welding mechanics at Carleton University, Canada. Since 2015, he has been an associate professor at Huazhong University of Science and Technology, China.

Bin Yi is a PhD student in the Department of Naval Architecture and Ocean Engineering, Huazhong University of Science and Technology, China. His research is focused on mitigation of welding-induced buckling with thermal tension and its application on thin plate section fabrication.

Qingya Zhang is a PhD student in the Department of Naval Architecture and Ocean Engineering, Huazhong University of Science and Technology, China. His research is focused on residual stress evaluation of thick plate structure with advanced measurements and a computational approach, as well as fracture performance assessment of the welded joint.

Computational Welding Mechanics for Engineering Application

Buckling Distortion of Thin Plate and Residual Stress of Thick Plate

Jiangchao Wang, Bin Yi, and Qingya Zhang

CRC Press is an imprint of the
Taylor & Francis Group, an **informa** business

Designed cover image: ©FOTO Eak

First edition published 2024
by CRC Press
2385 NW Executive Center Drive, Suite 320, Boca Raton FL 33431

and by CRC Press
4 Park Square, Milton Park, Abingdon, Oxon, OX14 4RN

CRC Press is an imprint of Taylor & Francis Group, LLC

This book is published with financial support from National Natural Science Foundation of China

Library of Congress Cataloging-in-Publication Data
Names: Wang, Jiangchao, 1983– author. | Yi, Bin, 1996– author. | Zhang, Qingya, 1990– author.
Title: Computational welding mechanics for engineering application : buckling distortion of
thin plate and residual stress of thick plate / Jiangchao Wang, Bin Yi and Qingya Zhang.
Description: First edition. | Boca Raton : CRC Press, 2024. |
Includes bibliographical references. |
Identifiers: LCCN 2023025767 (print) | LCCN 2023025768 (ebook) |
ISBN 9781032580722 (hardback) | ISBN 9781032580982 (paperback) |
ISBN 9781003442523 (ebook)
Subjects: LCSH: Welding–Data processing. | Buckling (Mechanics) | Residual stresses.
Classification: LCC TS227.2 .W36 2024 (print) | LCC TS227.2 (ebook) |
DDC 624.1/760285–dc23/eng/20231018
LC record available at https://lccn.loc.gov/2023025767
LC ebook record available at https://lccn.loc.gov/2023025768

ISBN: 978-1-032-58072-2 (hbk)
ISBN: 978-1-032-58098-2 (pbk)
ISBN: 978-1-003-44252-3 (ebk)

DOI: 10.1201/9781003442523

Typeset in Minion
by Newgen Publishing UK

Sincere Salute to

Professor Hidekazu Murakawa,
Osaka University, Japan

Professor John A. Goldak,
Carleton University, Canada

Contents

Acknowledgments

THE AUTHORS APPRECIATE FINANCIAL SUPPORT PROVIDED BY THE NATIONAL Natural Science Foundation of China (No. 51609091 and No. 52071151) and the High-Tech Ship Research Project: Development of Semisubmersible Lifting and Disassembly Platform (Ministry of Industry and Information Technology of the People's Republic of China: No. 614 [2017]), and also appreciate assistance in engineering application by China Merchants Heavy Industry (Jiangsu) Company and China Ocean Shipping – Kawasaki Ship Engineering Company for carrying out experiments and measurement.

CHAPTER 1

Introduction

With the advanced mechanical performance of high tensile strength steel, thin plate for lightweight fabrication of advanced ships and naval ships has been commonly employed, and thick plate for ocean engineering of offshore structures was also applied. Both thin plate and thick plate of high tensile strength steel are usually assembled by means of the fusion welding method due to its advanced joining features such as high productivity and flexible practice [1]; however, there are welding distortion and residual stress during the fusion welding process which could not be avoided.

Welding is the nondetachable joining or coating of components or base materials under the (mostly local) application of heat and/or pressure, with or without the use of filler materials [2]. In addition, welding distortion as well as residual stress is the result of rapid heating in the local region of welded joint and welded structure by fusion welding heat source such as welding arc, which have become both a research issue and an engineering topic. Welding distortion will seriously influence the fabrication accuracy; in particular, welding-induced buckling as a critical type of out-of-plane welding distortion will significantly influence the integrity of the thin plate structure and production schedule with straightening. Generally speaking, the welding-induced distortion of thin plate structures can be divided into in-plane and out-of-plane distortion. Welding shrinkage is the typical in-plane type. Out-of-plane welding distortion may be divided into two different kinds: bending distortion and buckling distortion. Figure 1.1

DOI: 10.1201/9781003442523-1

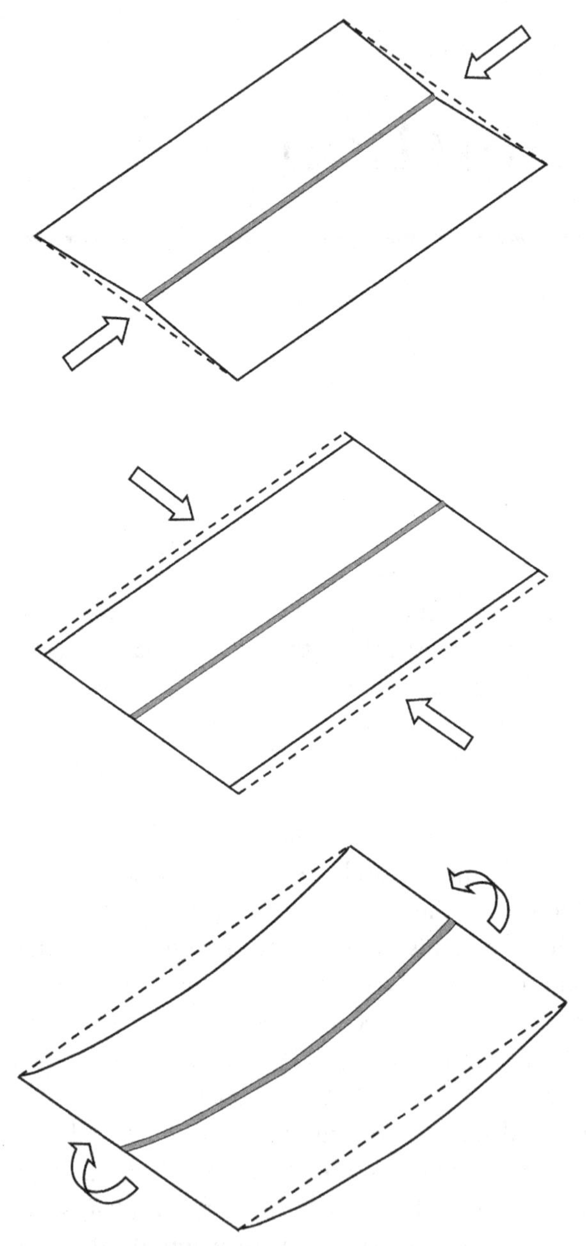

FIGURE 1.1 Basic types of welding distortion in thin plate butt-welded joint

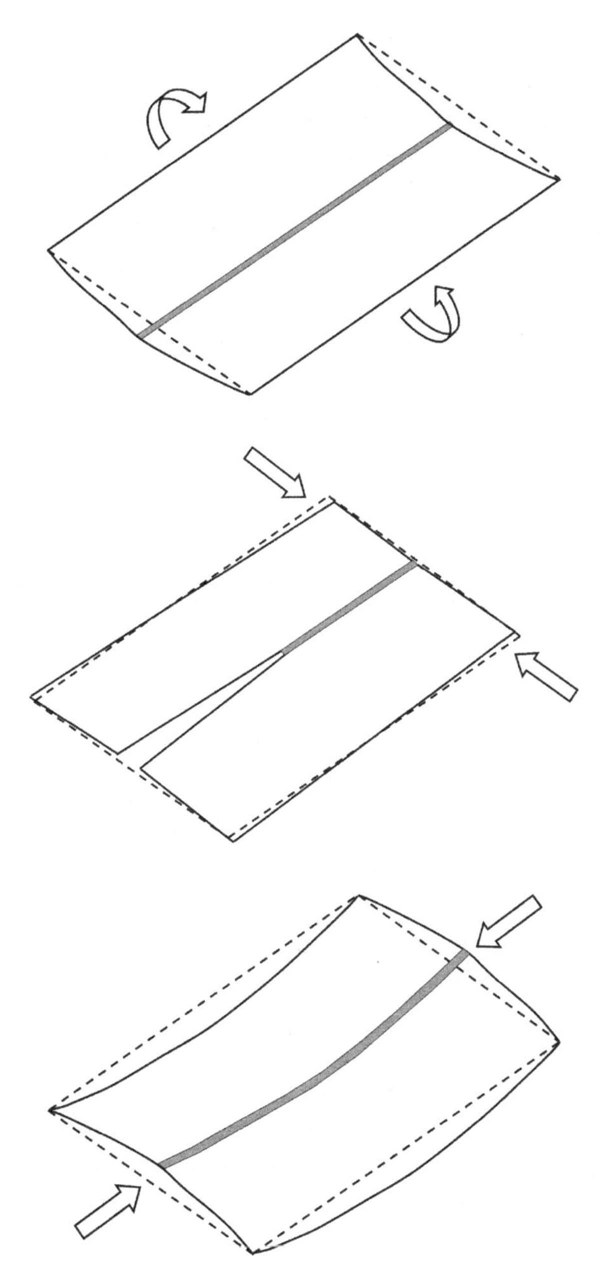

FIGURE 1.1 Continued

shows all typical welding distortions of a thin plate butt-welded joint. In particular, buckling distortion can take various modes.

Meanwhile, residual stress after cooling down will influence the balance of internal stress and the mechanical performance of welded structures such as fracture and fatigue performance in service life. Therefore, for the advanced manufacturing of ship vessel with thin plate of high tensile strength steel and offshore structure with thick plate of high tensile strength steel, it will be preferred to represent the corresponding mechanical behaviors based on the welding experiment and FE computation, while the generation mechanism of welding distortion and residual stress should be clarified and mitigation practice in the actual fabrication should also be proposed to enhance the welding quality with low residual stress and no welding distortion.

1.1 REVIEW OF COMPUTATIONAL WELDING MECHANICS

In the 1970s, Ueda and Yamakawa [3] were the first to present a computational approach to analyze welding thermal stresses and resulting residual stresses during welding using thermal elastic plastic FE analysis. In this method, the change of mechanical properties with temperature is considered. Butt- and fillet-welded joints are studied as examples. Many investigators including Hibbitt [4], Friedman [5], Lindgren and Karlsson [6], Matsubuchi [7], and Goldak [8] continued to develop the thermal elastic plastic FE analysis for the computation of temperatures, residual stresses, and distortions. Now thermal elastic plastic FE analysis is widely accepted to analyze welding problems for many types of welded joints and simple small-size welded structures.

Ueda and Murakawa [9] discussed the advances in computer technology and utilization of computers in welding research. They have shown that the progress in computer technology and numerical analysis has made it possible to simulate complex phenomena and to discuss welding problems from the quantitative aspect. They also pointed out that there are still some problems to be overcome before truly meaningful research can be achieved. Although finite element analysis has accomplished remarkable achievements in the investigation of short single-pass welds, welding is perhaps the most nonlinear problem encountered in structural mechanics [10]. Therefore, Lindgren [10–12] in 2001 published a review that presented the development of welding simulation from different aspects of finite element modeling and analysis, consisting of three parts: (1) increased complexity,

(2) improved material modeling, and (3) efficiency and integration. This review also has shown the possibilities in modeling and simulation of welding to encourage the increase in the use of the FEM in the field of welding mechanics. Meanwhile, the computational approach is clearly elaborated to be appropriate for those who are entering this field of research and useful as a reference for those already familiar with this subject. Then Lindgren [13] described the application of the finite element method as a tool of computational welding mechanics to predict thermal, material, and mechanical effects of welding and focused on modeling aspects.

As pioneers in the field of computational welding mechanics and the investigation of welding mechanical problems, such as temperature effects, strain, residual stress, and welding distortion, using the computational approach, Professor Ueda (Osaka University, Japan), Professor Goldak (Carleton University, Canada), and Professor Lindgren (Luleå University of Technology, Sweden) established the research disciplines of computational welding mechanics (CWM) as shown in Figure 1.2 [14].

A paper collection [15], titled "Computational Welding Mechanics", has been published in commemoration of the retirement of Professor Yukio Ueda from Osaka University in 1996, in which professor Ueda with his colleagues developed the basic theories and their applications in analysis, measurement, and prediction of welding residual stresses and deformation. Goldak and Akhlaghi [14] emphasized the exposition of computational principles and the application of computational welding mechanics to

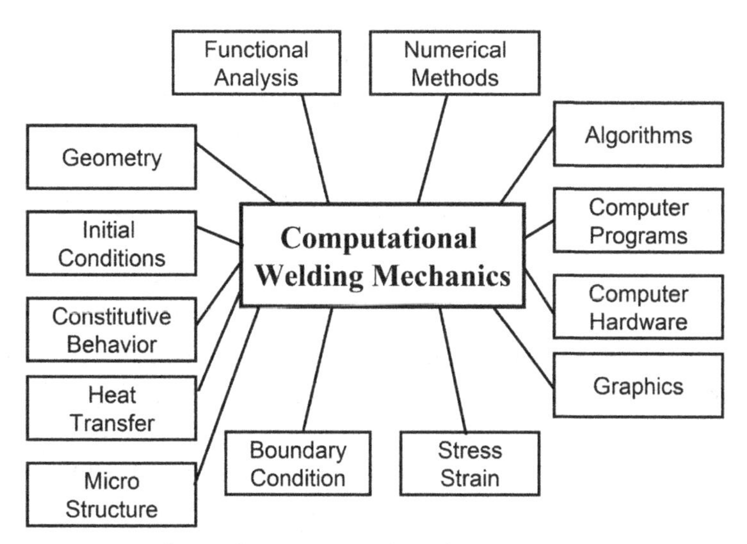

FIGURE 1.2 Disciplines of computational welding mechanics

problems in the industry. The primary aim of this work is to transfer technology from the research specialists who are developing it to the engineering society to apply it in design and production of welded structures. Goldak and Akhlaghi also pointed out that computational welding mechanics will be widely used in welding industry to give designers the capability to predict residual stress and welding distortion in welds and welded structures. Lindgren [16] in his monograph concluded that the aim of computational welding mechanics is to establish methods and models that are usable for control and design of welding processes to obtain appropriate mechanical performance of welded component or structure. This monograph is of benefit to anyone who wants to master CWM or simulation of manufacturing process in general.

1.2 EFFICIENT THERMAL ELASTIC PLASTIC FE COMPUTATION

Although thermal elastic plastic FE analysis provides good predictions of welding distortion, requirements on computer memory and computing time make it less applicable to large structures and products in actual engineering practice. In recent decades, more efficient computational methods, still based on the FE method, have been developed to investigate welding distortion of large structures. Many of these methods use shell FE models with elements much larger than those required in thermal elastic plastic FE analysis.

Michaleris and Debiccari [17] presented a numerical analysis technique combining 2D welding simulation with 3D structural analyses, which is effective when computer resources are limited. Bachorski et al. [18] presented a linear elastic finite element modeling technique called shrinkage volume method which can significantly reduce the solution time. Jung [19] employed plasticity-based distortion analysis to investigate the relationship between cumulative plastic strains and welding distortion. In this analysis Jung mapped each cumulative plastic strain component evaluated by a 3D thermal elastic plastic analysis incorporating the effects of moving heat and nonlinear material properties into elastic models using equivalent thermal strains. Camilleri et al. [20] examined the applicability of finite element analysis to the out-of-plane welding distortion. Transient thermal analysis was done on a 2D cross-section of the welded joint, and then the transverse angular deformation and the longitudinal contraction force are predicted. These results were then applied to a nonlinear elastic finite element model to provide predictions of the final overall deformations of butt-welded plates. Cheng [21] examined the in-plane shrinkage plastic strains through welding simulation of three types of simple welded joints. He found that these plastic strains are determined by the peak welding temperature and material's softening temperature range. Then an engineering approach was developed to introduce the obtained plastic strain into the FE model by means of applying equivalent thermal load.

Zhang et al. [22] developed a 3D "Applied Plastic Strain Method" to predict the welding distortion of structures. In this Applied Plastic Strain Method, six components of the plastic strain of each welding line are calculated by performing a thermal elastic plastic FE analysis assuming moving heat source on a small 3D model with a shorter length, and then the plastic strain components of the small models are mapped on a large 3D structural model to obtain the final distortion. Jung [23] carried out a shell-element-based elastic analysis for prediction of the distortion induced in ship panels. Only longitudinal and transverse plastic strains are employed to compute the welding distortion, without significant loss of accuracy compared with results of thermal elastic plastic FE analysis. Yang et al. [24] developed a stepwise computational method for prediction of welding distortion of large welded structures to reduce the computational time. This method consists of lump-pass modeling and mapping of plastic strain. The lump-pass modeling of local small models includes a thermal and mechanical analysis to obtain the plastic strain. The mapping plastic strain method is carried out to predict the welding distortion by mapping plastic strains evaluated from local small models into the global model.

Focusing on the improvement of computational speed, Nishikawa et al. [25, 26] developed the Iterative Substructure Method (ISM) to reduce the computational time in three-dimensional welding analyses. In this method, the computational domain is divided into a large quasilinear region and a small but moving strong nonlinear region [27]. Ikushima et al. [28, 29] developed a parallelized idealized explicit FEM using Graphics Processing Unit (GPU) to examine the mechanical behavior of large-scale welded structures, in which dynamic explicit FEM is employed using shorter computing time and less memory [30]. Runnemalm and Hyun [31] applied an adaptive strategy for coupled thermomechanical analysis of welding in order to reduce computer time. Parallel computation technique is also shown to be an effective method to improve computational speed. Kim et al. [32] developed a parallel multifrontal solver for finite element analysis of arc-welding process, in which phase evolution, heat transfer, and deformation of structures are considered. Tian et al. [33] performed a parallel calculation by establishing a computer cluster system composed of four PCs to predict the final distortion of a large complicated Al alloy structure during Electron Beam Welding (EBW).

Elastic FE analysis using shell element models and inherent strain and/or inherent deformation is proposed by a research group of Joining and Welding Research Institute at Osaka University, Japan, led by Prof Hidekazu Murakawa. This is a practical and efficient computational approach to investigate the welding distortion of large-scale and complex welded structures. Luo et al. [34] examined bead on plate as an ideal case of a butt-welded joint. In this work, inherent strain produced during the welding process is primarily determined by the highest temperature reached in the thermal cycles and the constraint at each point. Then an elastic FE analysis is carried out to predict the welding residual stress and distortion using the inherent strain as an initial strain. Murakawa et al. [35] proposed that the total distortion is determined as the elastic response of the structure to accumulated inelastic strains, such as plastic strain, creep strain, and that produced by phase transformation. The inelastic strains are defined as the inherent strains, which depend on local conditions and can therefore be computed using a limited number of typical joints and then used in different applications. They showed that inherent deformation can be evaluated by integrating the inherent strain distributed on each transverse cross-section and is easier to store in databases for prediction of welding distortion in the future. Luo et al. [36] first analyzed welding deformation of plates with longitudinal curvature using thermal elastic plastic FE analysis. Luo suggested that since both welding distortion and residual stress are produced by the inherent strain, then integration of the inherent strain gives not only the inherent deformations but also the tendon force. A method to predict the welding deformation by elastic FE analysis using the inherent deformations and the tendon force is proposed and its validity is demonstrated through numerical examples. Luo et al. [37], through a series of thermal elastic plastic FE analysis, established a database of inherent deformation for an aluminum alloy butt-welded joint. An elastic FE analysis based on this database was developed to predict the welding deformation for large welded structures such as a large aluminum alloy coping. Deng and Murakawa [38] developed a large deformation elastic FE analysis to accurately predict welding distortion. In this work, a thermal elastic plastic FE analysis is first used to evaluate the inherent deformations of various types of welded joints and an elastic FE analysis is performed to predict welding distortion based on the obtained inherent deformation. In this paper, the influence of initial gap on welding distortion is also investigated. Wang et al. [39] studied welding distortion of thin plates due to bead on plate welding with varying plate thickness using elastic analysis based on the inherent deformation method. They showed that this kind of elastic analysis using the inherent deformation not only saves a large amount of computing time but also preserves the computation accuracy as was verified by experimental results. Wang et al. [40] studied a spherical structure assembled from thin plates, in which the overall welding distortion due to each welding line is experimentally measured; then thermal elastic plastic FE analysis of the butt-welded joint is carried out to evaluate the inherent deformation. Then an elastic analysis utilizing the shell element model of this spherical structure and the evaluated inherent deformation is carried out to predict the welding distortion. Numerical predictions show good agreement with experimental measurements.

1.3 WELDING BUCKLING PREDICTION AND MITIGATION

In modern manufacturing, design of transport vehicles such as ships, automobiles, trains, and aircraft emphasizes minimizing weight to improve fuel economy and/or enhance the carrying capacity. Therefore the demand for lightweight structures assembled using thin plates densely stiffened has significantly increased. For thin plate ship structures, buckling-type distortion may be produced by the welding process. Buckling is considered to be the most critical type of welding distortion because of its instability and difficulty in straightening.

1.3.1 Prediction of Welding Buckling

Experimental investigation of welding-induced buckling can be found in the literature. Masubuchi [41] conducted experiments on buckling-type deformation of thin plates due to welding and presented the critical buckling wave length as influenced by the size of plate and welding conditions. Watanabe and Satoh [42] observed welding-induced buckling distortion, the so-called concave-convex or convex-concave type, in a series of experiments of bead on plate welding on thin plates. The critical buckling force and the moment when buckling occurs in welded plates with different aspect ratios is measured and studied [43]. Frank [44] indicated that thin plates can buckle during and after welding due to thermal and residual welding stresses. Critical residual buckling stresses are calculated for a number of different boundary conditions and a test program is conducted to partially verify the predictions. Terasaki et al. [45, 46] also investigated plate buckling behavior caused by welding, utilizing experiments and numerical analyses. They have shown that welding-induced buckling is affected by the welding conditions, size of plate, and material properties.

Taking advantage of numerical analysis and high performance computers, investigation of welding-induced buckling distortion using the computational approach is developed. Tsai et al. [47] studied the distortion mechanism and the effect of welding sequence on panel distortion. Using the finite element method, local plate bending and buckling were investigated. They concluded that, in their models, buckling does not occur in structures with a skin plate thickness of more than 1.6 mm unless the stiffening girder bends excessively. Tsai et al. [48] pointed out that the bifurcation phenomenon of buckling starts during the cooling cycle and this may continue to grow until the completion of the cooling process. An integrated experimental and numerical approach was applied to investigate the mechanism of welding-induced buckling evolution process. Meanwhile, eigenvalue analysis using welding longitudinal inherent strain distribution and 3D thermal elastic plastic FE analysis considering large deformation was performed on welded structural models to understand the buckling distortion process observed in the experiments.

P Michaleris [49, 50] pointed out that the compressive residual stress parallel to welding line contributes a loading that eventually results in buckling if the stress exceeds the critical buckling stress of the welded structure. He called this stress the Applied Weld Load (AWL), which is determined by performing local 3D thermal elastic plastic FE analysis of the welding process. A 2D shell model depicting the actual ship panel using a relatively coarse mesh is then used to perform an eigenvalue analysis that determines the various buckling modes caused by welding residual stress. The minimum eigenvalue obtained from the analysis is used to evaluate the resistance to buckling. This resistance to buckling has been called the Critical Buckling Load (CBL). A comparison of the AWL and the CBL indicates whether the welded structure is expected to buckle and what mode of buckling is likely to occur. Vanli and Michaleris [51] considered fillet welded joints with a particular emphasis on welding-induced buckling instabilities. The effects of stiffener geometry, welding sequence, welding heat input, and mechanical fixtures on the occurrence of buckling and the distortion pattern are investigated. Deo et al. [52] examined welding-induced buckling distortion by means of decoupled 2D and 3D approaches, where the investigation process is divided into two steps: first determination of welding residual stress based on a 2D thermal mechanical welding simulation, then obtaining the critical buckling stress and the buckling mode using a 3D eigenvalue analysis. Michaleris et al. [53] suggested that the magnitude of longitudinal residual stress is critical in the prediction of buckling distortion. An evaluation of modeling procedures to predict welding-induced buckling distortion incorporating a moving heat source, 2D and 3D small deformation analysis, 3D large deformation analysis, and 2D-3D applied plastic strain analyses is carried out by comparing computed residual stress and distortion with experimental measurements. Bhide et al. [54] compared submerged arc welding, gas metal arc welding, and friction stir welding in terms of their buckling propensity by measuring the longitudinal residual stress using the blind hole drilling method and welding distortion using digital gauges at points on the welded plates.

Mollicone et al. [55] studied buckling in the fabrication of large, thin plate welded structures and found that minor variations in fabrication procedures have significant effects on out-of-plane welding distortion. Buckling instability behavior is investigated considering the initial out-of-plane distortion due to tack welding and clamping conditions during welding and cooling. Huang et al. [56, 57] carried out a comprehensive assessment of fabrication technology of lightweight structures produced using relatively thin plates with a major initiative funded by the U.S. Navy Office of Naval Research. The assessment includes a comprehensive investigation of the fabrication and assembly processes of thin steel plate structures and their contributions to the final dimensional accuracy and distortion. An optical measurement system and advanced computational tools are employed to understand the underlying mechanism of buckling distortion and critical process parameters for ship panels. They noted that dimensional accuracy due to thermal cutting can have a significant impact on buckling distortion. They also noted that effective mitigation techniques for minimizing buckling distortion should either reduce the buckling driving force (fabrication-induced stresses) and/or increase the buckling resistance (panel geometric parameters). Yang and Dong [58] considered welding-induced buckling distortion as an unstable type of welding distortion and proposed a buckling analysis procedure incorporating welding-induced residual stress state.

Tajima et al. [59] pointed out that welding longitudinal shrinkage (tendon force) produces compressive stress in the surrounding plate fields that sometimes causes these plate fields to buckle. In order to avoid buckling, a series of thermal elastic plastic finite element analyses are carried out to predict welding tendon forces and transverse shrinkage/bending when utilizing continuous and intermittent welding with different welding specifications. A cross-stiffened panel of a car deck of a car carrier is investigated. Bidirectional residual stresses are then evaluated for different welding patterns. The effectiveness of welding patterns (continuous, parallel and zigzag intermittent welding) in reducing welding residual stress and preventing buckling is quantified. Deng and Murakawa [60] developed a prediction method for welding distortion, which combines the thermal elastic plastic finite element method and large deformation elastic FE method based on inherent strain/deformation method and the interface element. The inherent deformations of two typical welded joints used in a large thin plate structure were evaluated using thermal elastic plastic FE method. Then an elastic analysis using these inherent deformations is employed to investigate the influence of heat input, welding procedure, welding sequence, plate thickness, and spacing between the stiffeners on buckling propensity of this structure. Deng and Murakawa [61] developed a large deformation elastic FE analysis to accurately predict welding distortion. In this work, a thermal elastic plastic FE analysis was employed to evaluate the inherent deformations of various types of welded joints beforehand, and an elastic FE analysis was then performed to predict welding distortion based on the obtained inherent deformation. Concentrated on the engineering application for accurate fabrication with predication and reduction of welding distortion, typically welded joints in the considered ultralarge container ship [62], jack-up rig offshore [63], and semisubmersible lifting and disassembly platform [64] were examined by means of experiment or computation for evaluation of welding inherent deformation. Then, elastic FE computations were carried out for welding distortion predictions of actual welded structures, optimization of welding sequence, and application of inverse deformation were also examined for welding distortion reduction.

In addition, Wang et al. [65] examined bead-on-plate welding with 2.28 mm thin plates experimentally to investigate welding-induced buckling. Considering large deformation and initial deflection, an elastic FE analysis with welding inherent deformation was then carried out to predict out-of-plane welding distortion, which showed a good agreement with measured distortion. With eigenvalue analysis, the generation mechanism of buckling was clarified as follows: tendon force (longitudinal inherent shrinkage) is the dominant reason to produce buckling and the disturbance (initial deflection or inherent bending) triggers buckling but does not influence the buckling mode. Later, Wang et al. [66] conducted a welding experiment on thin plate stiffened structure, and a large twisting distortion was observed after cooling down, and its out-of-plane welding distortion was numerically examined as welding buckling response by means of a thermal elastic plastic FE analysis and an elastic FE analysis based on the concept of welding inherent deformation. Predicted out-of-plane welding distortions by both computational methods show that not only the deformed shape but also the magnitudes have good agreement compared with experimental measurement. For actual ship panel with thin plates, Wang et al. [67–71] investigated welding-induced buckling with an elastic FE analysis based on the inherent deformation method, while large deformation theory is necessary to consider behavior of geometrical nonlinearity. They also point out that out-of-plane welding distortion is eventually determined by bending deformation and buckling deformation together, and characteristics of welding-induced buckling may not be observed due to much larger magnitude of bending deformation.

1.3.2 Mitigation of Welding Buckling

At present, it is impossible to completely eliminate or correct welding distortion because of the difficulty due to the nonlinear, irreversible nature of the welding and the straightening processes. However, welding distortion should and can be minimized to an acceptable magnitude to control dimensional tolerances during assembly.

Several mitigation techniques of welding-induced buckling, such as external restraint [72, 73], laser welding [74], intermittent welding [75], rolling [76], mechanical correction [77], electromagnetic impact [78], and flame straightening after welding [79, 80], were reported by researchers. Mitigation practice during the design stage has its advantage for reduction of welding-induced buckling, which is also limited due to the dimension of welded structure, consideration of structural strength, and the possibility of fabrication processing. Thus, mitigation practice during fabrication process, with its efficient and practical features, is usually considered and employed to reduce the welding buckling. Local temperature control

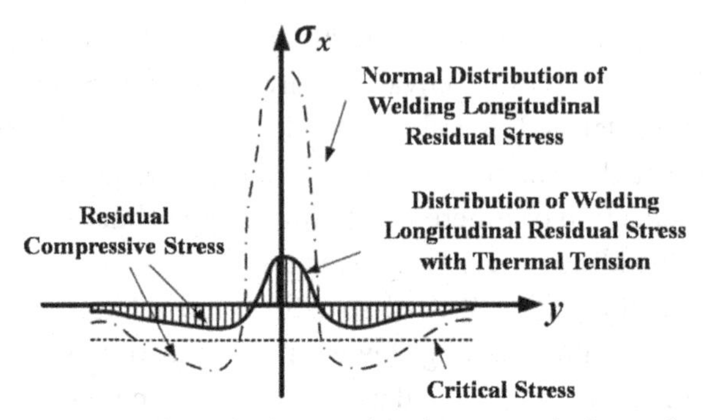

FIGURE 1.3 Variation of residual stress distribution with thermal tension

approaches [81–83], in particular, thermal tension technique with an additional heat source, are widely examined and employed in shipbuilding due to their practical and effective features. In principle, since welding-induced buckling is complicated and difficult to eliminate during postprocess straightening stage, the mitigation techniques applied during design and in-process stages are preferred.

Thermal tensioning, first applied by Burak et al. [84] and later patented by Guan [85, 86], describes a group of in-process methods to control welding distortion. In these methods, local heating and/or cooling strategies are applied during the welding process. As shown in Figure 1.3, welding residual stress will be redistributed with the effect of the thermal tension of additional heating, and longitudinal compressive residual stress will be less than the critical compressive stress to avoid the occurrence of welding-induced buckling.

Thermal tensioning using additional heating sources can be classified into two types; one is static thermal tensioning (STT, additional heating and cooling sources are applied on the whole weld), and the other is transient thermal tensioning (TTT, additional heating sources travel with the welding gun). STT treatment is impractical for fabricating large-scale or complicated welded structures due to the high cost and complex equipment. However, TTT treatment is commonly employed to control the welding distortion in real time for its high productivity and simple operation.

In order to minimize welding-induced bucking of thin section structure, Michaleris et al. [87] examined a new technique, so-called steady thermal tensioning (STT) as demonstrated in Figure 1.4. In detail, a temperature differential was obtained by cooling the weld region with impingement

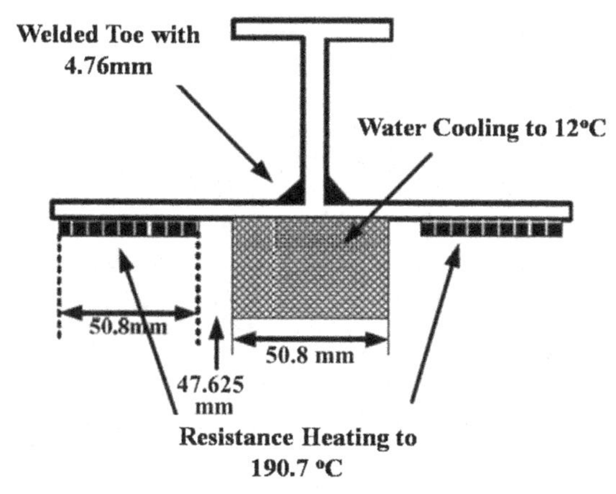

FIGURE 1.4 Image of steady thermal tensioning with impingement tap water and resistive heating blankets

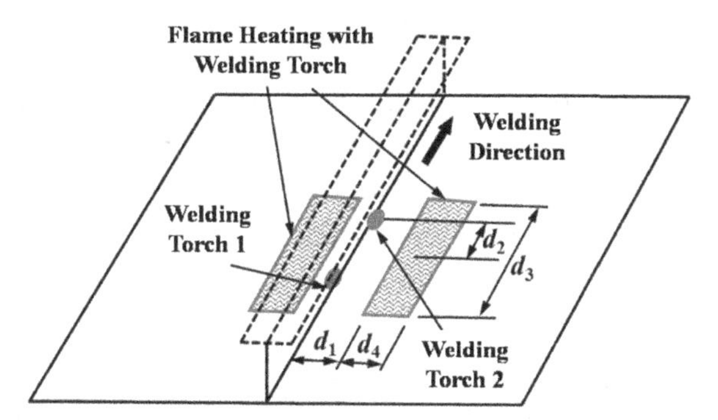

FIGURE 1.5 Image of transient thermal tensioning for buckling mitigation

tap water below the weld, and heating by the plate with resistive heating blankets. Later, Deo et al. [88] conducted the experimental verification and demonstration of TTT treatment to eliminate welding-induced buckling as demonstrated in Figure 1.5. The results indicated that welding-induced buckling results from compressive stresses at the free edges of the panel exceeded the critical buckling stress, and can be eliminated with required tensile stress produced by transient thermal tensioning (TTT) treatment. Yang et al. [89] developed finite element models for TTT treatment of complex thin steel panels, which were demonstrated to be effective in reducing the propensity of buckling distortion. Huang et al. [90] implemented a series of experiments with transient thermal tension for welding-induced buckling mitigation. In addition, magnitude of compressive residual stress

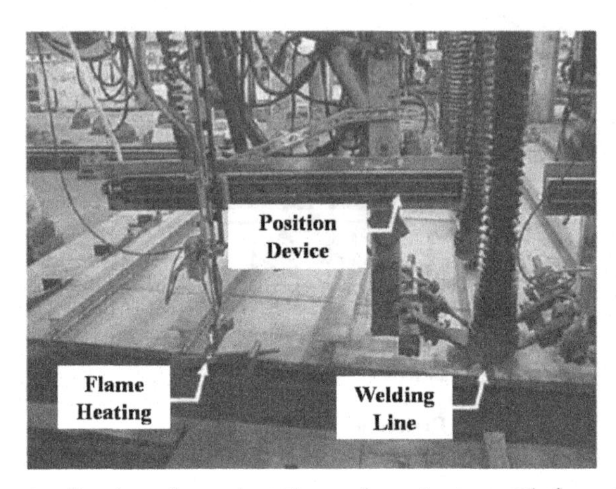

FIGURE 1.6 Application of transient thermal tensioning with flame heating

away from welding line was modified, and welding-induced buckling can be avoided when compressive residual stress is less than a critical condition determined by structural stiffness. Advantages of transient thermal tensioning (TTT) treatment in manufacturing lightweight structures as demonstrated in Figure 1.6 were summarized with high production efficiency, cost-saving, and no additional correction works.

Souto et al. [91] examined the influence of the position of auxiliary torches for transient thermal tensioning (TTT) treatment in T-joints. The results showed that the torches should be used over the neutral line for reducing the out-of-plane distortion, and TTT treatment does not have any impact on the maximum temperature and stress on the weld. Moreover, these new compressive stresses from the auxiliary torches can compensate for the force during the welding process, leading to the improvement of distortion at the end of the cooling. Zhang et al. [92] proposed a new electron beam welding method with simultaneous multibeam preheating on both sides of the weld, and reduction by 80% of the buckling distortion was achieved. Li et al. [93] performed a hybrid technology of transient thermal tensioning and trailing intensive cooling for controlling welding residual stress and distortion. In their research, temperature field and longitudinal residual stress were adjusted by this hybrid technology, and the reductions of the welding residual stress and distortion of joint reached 65% and 58%, respectively. Based on welding inherent strain, Yi et al. [94] considered the buckling deformation during thin plate butt welding of 304

L stainless steel, while computed out-of-plane welding distortion has good agreement with measurement. Two symmetrical additional heat sources located at the region off welding line with 110 mm and moving synchronously with welding arc were numerically examined. In mechanism, transient thermal tension will weaken self-constraint in longitudinal direction supported by base materials, decrease magnitude of longitudinal shrinkage force, and eventually reduce welding-induced buckling. Typical fillet-welded joints in considered cabin structures were numerically examined beforehand, while thermal and mechanical behaviors were represented with nonlinear thermal elastic plastic FE computation [95]. Then, welding inherent deformation was proposed as driving force for welding distortion prediction, which can be evaluated by integration of residual plastic strain for each typical welded joint. Elastic FE computation with welding inherent deformation as mechanical load was carried out to predict out-of-plane welding distortion of considered cabin structure, while cases of conventional welding and welding with transient thermal tension were examined. Better fabrication accuracy can be obtained with application of transient thermal tension due to less welding inherent deformation. In addition, out-of-plane welding distortion is determined by initial deflection, welding buckling and bending deformation, and features of welding-induced buckling may not be observed due to large magnitude of welding bending deformation.

Welding deformation due to a bead-on-plate welding under a non-constraint-free condition and a jig constraint condition was investigated by experiment and transient thermal elastic plastic FE analysis while welding angular distortion has been greatly reduced by the jig constraint [96]. Wang et al. [97] investigated welding-induced buckling with an elastic analysis based on the inherent deformation method. They considered a thin plate welded structure, for which the inherent deformation is evaluated for a typical welded joint by means of 3D thermal elastic plastic FE analysis. Straightening using line heating on the plate side opposite to the stiffener side to reduce buckling distortion is also investigated utilizing the same approach. Pazooki et al. [98] studied transient thermal tensioning (TTT) treatment during the butt joint with 2 mm thick DP600 steels by numerical and experimental approaches. It was found that transient thermal tensioning (TTT) treatment can successfully reduce out-of-plane distortion; the closer the burners were to the welding line, the higher distortion was obtained.

1.4 MEASUREMENT AND PREDICTION OF WELDING RESIDUAL STRESS

With the application of thick plate with high tensile strength steel in the fabrication of ship and offshore structure, pressure vessel, bridge, and so on, welding residual stress due to its complexity and importance is becoming an issue of engineering and research. In order to evaluate the welding residual stress, there are lots of measurement approaches to examine the magnitude and distribution of residual stress in the welded joint and welded structure [99].

In general, the measurement approach of welding residual stress can be divided into two types: destructive and nondestructive approaches. There are some measurements of the destructive approach such as little hole method, contour method, and so on, while measurement of nondestructive approach includes X-ray diffraction method, neutron diffraction method, and ultrasonic method.

Measurement approach of welding residual stress of nondestructive approach requires high surface smoothness of testing specimen, and welding residual stress near surface of testing specimen could only be measured. As shown in Figure 1.7, the comparison of measured accuracy

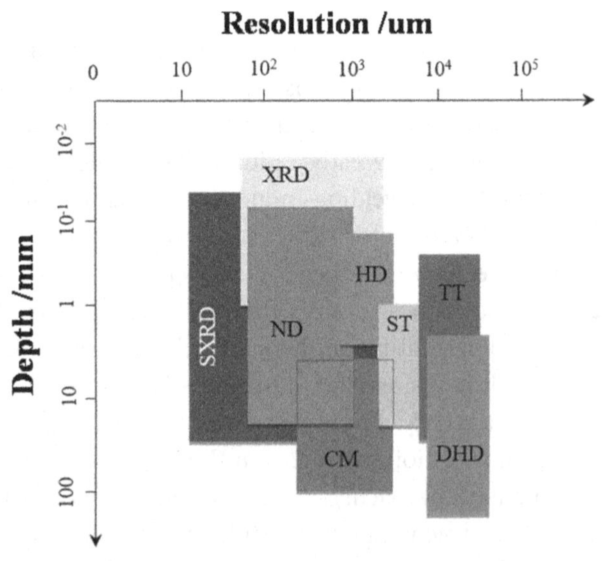

FIGURE 1.7 Comparison of measured accuracy and depth with different measurement approaches of welding residual stress

and depth with different measurement approaches of welding residual stress was demonstrated. It can be seen that X-ray diffraction method has a high measured accuracy and limitation in the measured depth, little hole method with best measured depth has less measured accuracy. Contour method has both advantages with high measured accuracy and well-measured depth, which is widely employed for welding residual stress measurement of welded joint with thick plate.

Woo et al. [100–102] employed the neutron diffraction method, little hole method, and contour method to evaluate welding residual stress of butt-welded joint with high tensile strength steel and thickness of 50–70 mm, while distribution tendence and magnitude of welding residual stress obtained with different measurement approach are similar. Balakrishnan et al. [103] fabricated pressure vessel with different welding methods beforehand, and heat treatment was then carried out to relax the welding residual stress of examined pressure vessel. Contour method and neutron diffraction method were both employed to measure the welding residual stress of examined pressure vessel and that after heat treatment. Measured welding residual stress by means of contour method and neutron diffraction method is similar, and heat treatment has a significant effect on reducing the welding residual tensile stress. Vasileiou et al. [104] conducted the butt welded joint of high tensile strength steel with thickness of 130 mm by means of narrow gap TIG, narrow gap SAW, and electron beam welding, while contour method and little hole method were both employed to measure the welding residual stress of examined welded joint. Maximal magnitude of welding tensile stress occurs in the region of welded zone by means of arc welding method, while the maximum magnitude of welding tensile stress occurs in the region near welding zone by means of electron beam welding.

For the ferrite steel with thickness of 80 mm, Kartal et al. [105] obtained the butt welded joint of thick plate by means of multipass welding and single pass welding, and different measurement approaches were employed to evaluate the influence of welding sequence and plate size on the distribution of welding residual stress, while the contour method is a practical measurement approach to obtain welding residual stress of welded joint with thick plate. Park et al. [106] concentrated on the welding residual stress of welded joint of high tensile strength steel with thickness of 70 mm, and inherent strain method was employed to measure the welding residual stress while considering the influence of initial rolling stress. Measured result of the neutron diffraction method confirmed the accuracy of residual stress measurement with inherent strain method, and the residual tensile stress in the region of base material is caused by initial rolling stress. Qiang et al. [107] employed the little hole method to measure the welding residual stress of butt-welded joint of Q345 with the plate thickness of 16 mm, 40 mm, 60 mm, 80 mm, and 100 mm. When the plate thickness is 16 mm, welding residual stress has a uniform distribution in the thickness direction. Distribution of welding residual stress in the thickness direction will be nonuniform when the plate thickness equals 40 mm. When the plate thickness is larger than 60 mm, distributions of longitudinal and transverse residual stress in the thickness direction will be the "C" form. In order to obtain the welding residual stress of welded joint of high tensile strength steel (SAF2205) with thickness of 55 mm, Jiang et al. [108] developed a novel neutron diffraction method, while its measured welding residual stress has a good agreement comparing with that by means of contour method.

Due to the limitation of measurement approach of welding residual stress such as experimental cost, measured time consumption, and measured accuracy influenced by operation procedure and equipment, transient thermal elastic plastic FE computation is then presented and increasingly employed to evaluate the welding residual stress, while the generation and evolution of residual stress during welding process can also be examined.

With the development of high-performance computer and computational approach, Ueda and Yamakawa [3] firstly proposed thermal elastic plastic FE formulation to examine the welding temperature, mechanical response, residual stress, and welding distortion considering the temperature-dependent material properties In 1971. Smith et al. [109] measured and predicted the welding residual stress of butt welded joint of high tensile strength steel (A533B) with thick plate, which was fabricated by electron beam welding. Computed welding residual stress has a good agreement compared with measurement results and predicted maximal magnitude of tensile residual stress is a little bit higher than that of measurement. Hwang et al. [110] employed the X-ray diffraction method and FE computation to examine the welding residual stress of EH40 welded joint with plate thickness of 80 mm, while computed welding residual stress well agrees with the measured results.

Wan et al. [111] examined the welding residual stress of Q345 butt-welded joint with plate thickness of 32 mm by means of contour method,

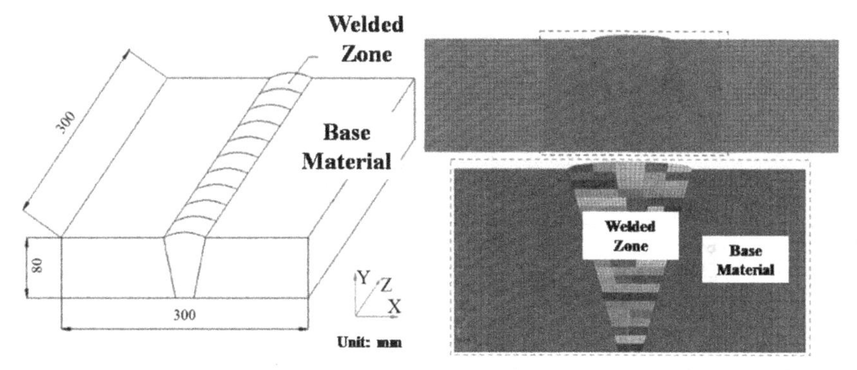

FIGURE 1.8 Dimension and FE mesh of EH47 butt-welded joint with plate thickness of 80 mm

neutron diffraction method, and FE computation, while the distribution of longitudinal welding residual stress in the transverse direction with "M" form could be observed. Later, Jiang et al. [112] examined the welding residual stress of EH47 butt-welded joint with plate thickness of 80 mm by means of neutron diffraction method and FE computation. As shown in Figure 1.8, FE model and mesh of examined EH47 butt-welded joint with plate thickness of 80 mm were demonstrated, and good agreement between predicted welding residual stress and measured results was obtained. Zhou et al. [113] employed thermal elastic plastic FE computation to predict the welding residual stress of EH47 multipass butt-welded joint with plate thickness of 70 mm, and computed results were validated with measured welding residual stress using little hole method.

Meanwhile, due to the thick plate of welded joint with high tensile strength steel, the FE mesh with massive node and element will be employed for conventional thermal elastic plastic FE analysis, which will require huge computer resources as well as computing time. In order to consider both computing time and computed accuracy, Murakawa et al. [114] proposed an advanced computational approach with iterative substructure method (ISM), while computing domain is divided into strong nonlinear region and weak nonlinear region (linear region). In general, welding line and its nearby region are all considered to be strong nonlinear domain with fine mesh, and the base material region far away from the welding line is considered to be weak nonlinear region as well as linear region with coarse mesh. For the lower computational efficiency of welded joint with thick plate, Ikushima and Shibahara[115] developed an explicit FE computation with parallel computation based on GPU. As shown in

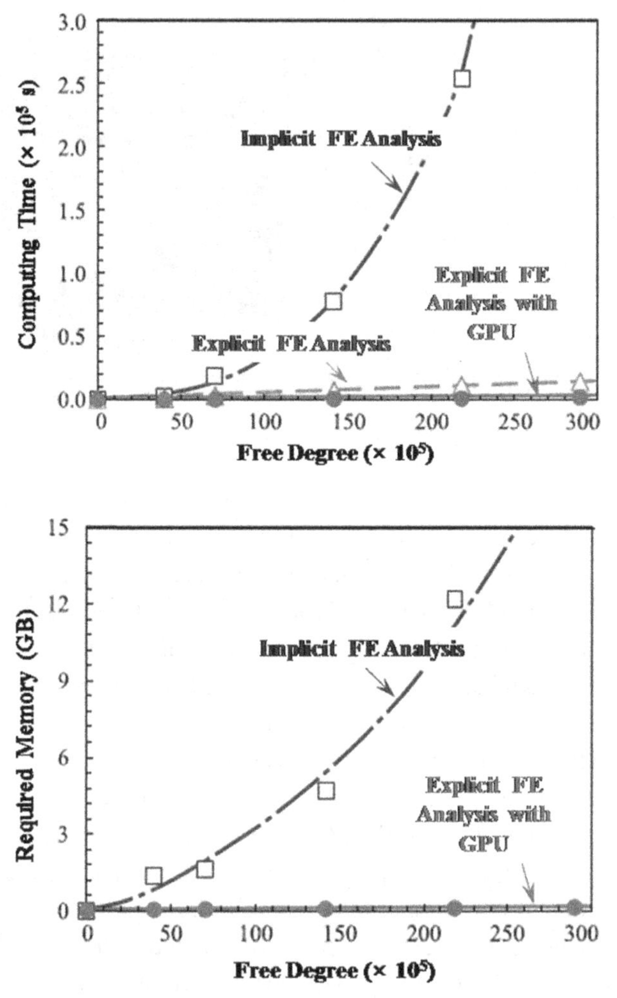

FIGURE 1.9 Comparison of computing time and computational memory with different FE computation

Figure 1.9, with the increasing free degree of FE model of examined welded joints, conventional implicit FE computation will consume massive computing time and computer resources, while explicit FE computation with parallel computation based on GPU could significantly reduce the requirement for computer resources and improve computational efficiency. Deng et al. [116] examined the generation and evolution of welding residual stress of multipass butt-welded joint by FE computation with moving heat source, transient heat source, and fixed heat source, while transient heat source could be employed to significantly reduce computing time without loss of computed accuracy.

Furthermore, during the welding process of high tensile strength steel, welding residual stress is generated due to the nonuniform transient heating and cooling in the local region, which magnitude is also influenced by welding metallurgical reactions such as solid phase transformation as well as microstructure evolution. In addition, thick plate is usually produced by means of thermal machinery rolling process (TMCP) with complicated internal stress before welding, which will more or less influence the eventual welding residual stress during FE computation. In order to improve the computed accuracy of welding residual stress, Ren et al. [117] proposed a 3D welding FE formula with thermal, metallurgical, and mechanical to examine the temperature, microstructure, and welding residual stress of butt-welded joint with high tensile strength steel based on the commercial software of SYSWELD. The computed welding residual stress has a good agreement compared with measurement results. Based on the parallel computation and iterative substructure method (ISM) with considering solid phase transformation and temperature-dependent material properties, Ma et al. [118] developed an advanced FE computation to predict the welding residual stress of flash butt-welded joint of high train track (U71Mn), while the solid phase transformation significantly influences the distribution and magnitude of welding residual stress of welded joint of rail steel. Jiang et al. [119] employed the measurement and FE computation to examine the welding residual stress of EH40 butt-welded joint with plate thickness of 25 mm. And volume expansion of Martensite phase transformation during the cooling process will significantly reduce the peak magnitude of longitudinal tensile stress. Park et al. [120] employed thermal elastic plastic FE computation to examine the influence of initial rolling stress on welding residual stress of butt-welded joint of high tensile strength steel with plate thickness of 70 mm. It is concluded that initial rolling stress is the dominant cause of influencing distribution of welding residual stress in the region of base material of butt welded joint of high tensile strength steel. Considering the influence of Bainite phase transformation during cooling process, Xu et al. [121] developed a FE computation with thermal, metallurgical, and mechanical analysis to predict the welding residual stress of welded joint with high tensile strength steel. Good agreement between computed and measured results shows that Bainite phase transformation can significantly reduce the magnitude of longitudinal tensile stress in the region of welding line.

Therefore, there is a lot of literature on measurement and prediction of welding residual stress of welded joint of high-tensile strength steel with thick plate. Meanwhile, lots of advanced techniques such as parallel computation, iterative substructure method (ISM), solid phase transformation, and initial rolling stress, were proposed and employed to enhance the computational efficiency and computed accuracy.

1.5 FRACTURE PERFORMANCE OF WELDED JOINT

Fracture performance of high tensile strength steel as well as its welded joint is an essential research issue and engineering problem. Theory about stress intensity factor (SIF), J integral, and crack tip opening displacement (CTOD) is widely employed to evaluate the fracture performance of welded joint during the service time. Meanwhile, initialization and growth of micro defects such as void of metal material is usually considered to be the elementary cause to generate fracture for high tensile strength steel and its welded joint.

Gurson [122] established the relation between void evolution of material and material yielding, and proposed the Gurson damage model and its corresponding yield function. Later, Tvergaard [123, 124] considered the interaction between microvoids, and presented parameters of q_1, q_2, and q_3 to modify the yield function of Gurson damage model, which could be accurately examine the growth process of void. Then, Chu and Nedleman [125] presented the parameters (ε_N, S_N, f_N) about void initialization to improve the void initialization criterion. Meanwhile, the critical void volume fraction (f critical) and fracture void volume fraction (f fracture) were also presented to examine the void aggregation with the modification of GTN damage model. There are lots of applications with GTN damage model for theory research and engineering solutions. In addition, GTN damage model usually has five basic parameters (q_1, q_2, q_3, ε nucleation, and S nucleation) with the constant value, and four advanced parameters (f0, f critical, f fracture, f nucleation) confirmed by high tensile strength steel and its material defects, which have significant influence on fracture performance of high tensile strength steel and its welded joint. Thus, parameter value of GTN damage model of high tensile strength steel should be confirmed by means of experiment and FE computation beforehand, and then fracture performance of high tensile strength steel and its welded joint can be examined and predicted.

Considering the influence of shear loading, Achouri et al. [126] proposed an extended GTN damage model to predict the fracture toughness of high tensile strength steel, parameter value of GTN damage model was presented, and computed results about fracture performance has a good agreement comparing with experimental results. Based on GTN damage model, Rakin et al. [127] examined the influence of width of welded joint, prefabricate crack length, and FE mesh on fracture toughness of welded joint with high tensile strength steel and combination of experiment and FE computation was then employed to analyze fracture behavior of welded joint with high tensile strength steel. Chhibberet al. [128] examined the influence of parameters of GTN damage model on mechanical performance of high tensile strength steel, and fracture experiment of pipeline was carried out to validate the reliability of GTN damage model. Carlos et al. [129] considered the plastic anisotropy, mix isotropic hardening, void nucleation–propagation–aggregation, and proposed an extended GTN damage model to predict the fracture behavior of high tensile strength steel, while its parameters were confirmed by means of optimization analysis, sensitivity analysis, and a series of experiment.

In order to predict the fracture behavior of high tensile strength steel with different stress configurations, Marteleur et al. [130] considered the anisotropy of failure and then proposed an extended GTN damage model to examine the influence of stress triaxial degree and LODE angle on void initialization as well as fracture strength. Qiang and Wang [131] employed the FE computation based on GTN damage model to examine crack growth in the different regions of girth welded joint of X80 pipeline steel and parameters of GTN damage model were also confirmed. Wu et al. [132] predicted the mechanical response of girth welded joint with X80 pipeline steel and Q235 low carbon steel by means of GTN damage model. Computed true stress-strain curve of examined girth welded joint has a good agreement comparing the experiment results.

In order to examine the influence of welding residual stress on fracture performance of welded joint, Gadallah et al. [133] developed a computation approach of stress intensity factor (SIF) based on field of welding residual stress and evaluated the stress intensity factor (SIF) of butt-welded

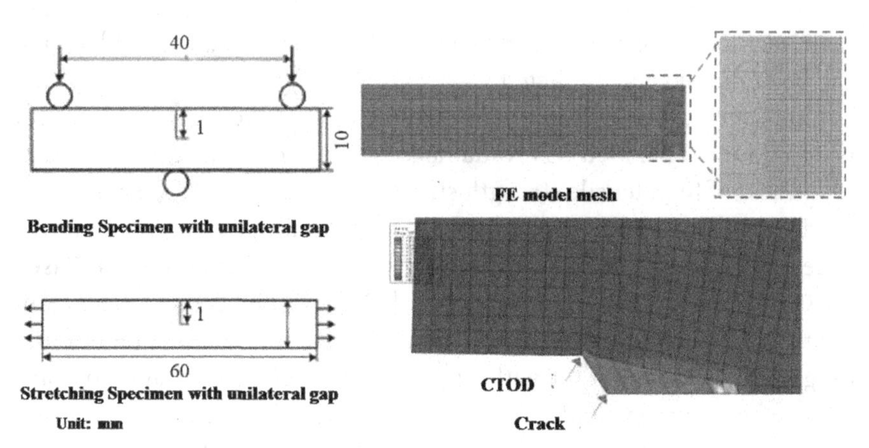

FIGURE 1.10 FE mesh model of fracture testing specimen of pipeline

joint considering welding residual stress. For the welded joint of pipeline, Lee et al. [134] employed thermal elastic plastic FE computation to obtain the welding residual stress beforehand, and approach of weight function was then employed to evaluate the influence of welding residual stress on the stress intensity factor (SIF) of welded joint. Pan et al. [135] employed different heat source models of welding to examine thermal and mechanical response of box-beam-column welded structures and examined the influence of welding residual stress on fracture behavior of welded structure. Welding residual stress will result in increasing equilibrium equivalent plastic strain in the region of welded zone and cause fracture failure of welded joint. As shown in Figure 1.10, Zhuo et al. [136] employed commercial software of ABAQUS with GTN damage model to analyze the influence of welding residual stress on fracture toughness of pipeline steel. FE mesh model of specimen of bending and stretching of unilateral gap was made and used to consider the influence of welding residual stress by applying inherent strain. Ren et al. [137] took the welding residual stress as initial stress of FE model and employed GTN damage model to predict the influence of welding residual stress on the processes of void initialization, propagation, and aggregation. It was also pointed out that tensile residual stress will significantly reduce the resistance of crack propagation, and the influence of welding residual stress on resistance of crack propagation will be weak with the crack propagation.

1.6 SUMMARY

The research progress about welding distortion, in particular welding-induced buckling of thin plates welding, and residual stress about thick plate with high tensile strength steel, was holistically reviewed. In addition, the advanced measurement method, FE computation approach, mitigation practice of welding-induced buckling, and evaluation approaches of welding residual stress and fracture performance were all well introduced.

CHAPTER 2

Elementary Theories and Methods

FOR THE WELDING MECHANICS investigation with computational analysis, there are lots of elementary theories as well as methods such as transient thermal elastic plastic FE, welding inherent deformation, elastic stability of welding buckling, solid phase transformation during welding, welding residual stress evaluation, mechanical performance assessment of welded joint, and so on.

2.1 TRANSIENT THERMAL ELASTIC PLASTIC FE ANALYSIS

In the TEP FE analysis, two processes are considered: the thermal process and the mechanical process. Despite the fact that the thermal process has a decisive effect on the mechanical process, the mechanical process has only a small influence on the thermal process, and a coupled thermal-mechanical analysis is not necessary. Therefore, thermal-mechanical behavior during welding is analyzed using uncoupled thermal-mechanical formulation. This uncoupled formulation considers the contribution of the transient temperature field to stress through thermal expansion, as well as temperature-dependent thermal-physical and mechanical properties. The solution procedure consists of two steps. First, the temperature distribution history is computed using heat transfer analysis. Then, the obtained transient temperature distribution is employed as a thermal load in the

DOI: 10.1201/9781003442523-2

subsequent mechanical analysis, in which residual stresses, plastic strains, and displacements are computed.

The main TEP computational model for transient thermal and stress analysis is described as follows:

2.1.1 Thermal Analysis

A welding employs a moving local high-intensity power source to the part which generates a sharp thermal profile in the weld pool, heat-affected zone (HAZ), and base metal. The 3D transient temperature is basically determined by solving the partial differential equation for the conservation of energy on a domain defined by a FEM mesh in a Lagrangian formulation.

$$\begin{aligned} &\dot{h} + \nabla q + Q = 0 \\ &q = -\kappa \nabla T \\ &\dot{h} = \rho c \frac{dT}{dt} \end{aligned} \tag{2.1}$$

where h is the specific enthalpy and the superimposed dot denotes the derivative with respect to time, T is the temperature and ∇T is the temperature gradient, Q is the power per unit volume or the power density distribution, κ is the thermal conductivity, c is thermal capacity, and ρ is density. These material properties are usually temperature dependent.

In this analysis, the initial temperature is often assumed to be room temperature, and body heat source with uniform power density Q (w/m^3: welding arc energy/volume of body heat source) is employed to model the heat source of welding arc, the value of which is determined by heat input as well as welding condition as defined in Eq.(2.2).

$$Q = \frac{\eta \times U \times I}{v} \tag{2.2}$$

where η means thermal efficiency of welding arc; U and I are welding voltage and welding current, respectively; and v means welding velocity.

Besides considering the moving heating source, heat losses due to convection and radiation are also taken into account in the finite element model. A combined convection and radiation boundary condition generates a boundary flux q (w/m^2) on all external surfaces. This flux is determined by heat convection and radiation together from the following equations:

$$\begin{aligned} q &= q_c + q_r \\ &= h(T - T_{ambient}) + \varepsilon C[(T+273)^4 - (T_{ambient}+273)^4] \\ &= [h + \varepsilon C[(T+273)^2 + (T_{ambient}+273)^2][(T+273) \\ &\quad + (T_{ambient}+273)]](T - T_{ambient}) \end{aligned} \tag{2.3}$$

where qc and qr are heat convection and radiation, respectively; h is coefficient of heat convection; ε is emissivity of the object which will be equal to 1 for ideal radiator; and C is the Stefan-Boltzmann constant.

Meanwhile, the FEM domain is dynamic, which will change with filler metal added to the welding pass at each time step during welding. After welding pass is completed, the time step length is increased exponentially until sum of computed time reaches a defined maximum examined time, and the analysis halts.

2.1.2 Stress Analysis

Given the density ρ, the fourth-order elastic plastic tensor D being a 6 × 6 matrix, the body force b, and the Green-Lagrange strain ε, stress analysis solves the conservation of momentum equation at the end of each time step that can be written in the form of Eq. (2.4) in which inertial forces are neglected.

$$\begin{aligned} &\nabla\sigma + b = 0 \\ &\sigma = D\varepsilon \\ &\varepsilon = \frac{\nabla u + (\nabla u)^T + (\nabla u)^T \nabla u}{2} \end{aligned} \tag{2.4}$$

The initial stress often is assumed to be stress free. The system is solved using a time marching scheme with time step lengths with certain magnitude during welding and usually an exponentially increasing time step length after welding has stopped. In terms of boundary conditions, the part is free to deform but rigid body motions are fixed as constraint.

2.2 THEORY OF PLASTICITY

When welded structures are subjected to excessively compressive thermal stress due to arc heating, weld joints may undergo plastic strain and plastic deformation. Plastic deformation is caused by the motion of dislocations, and dislocations are mainly driven by shear stress and hydrostatic pressure does not influence its motion; thus the yield condition of a metal is independent of hydrostatic pressure unless it is extremely large. Theory of plasticity with incremental form for stress-strain relation has been employed to solve problems involving unloading in most finite element methods. The primary research fields of plasticity are flow theory, yield condition, and strain hardening rule, which will be discussed respectively in the following.

2.2.1 J_2 Flow Theory

As mentioned above, the flow theory is superior to the deformation theory for numerical analysis of practical problems. In this study, J_2 flow theory is employed for FE computation, and its details are discussed in follows.

Since hydrostatic component of the stress tensor has no effect on the yielding and the plastic deformation of the material, the plastic phenomenon is governed only by deviatoric stress S_{ij}, which is defined by

$$\begin{aligned} S_{ij} &= \sigma_{ij} - p \\ p &= \frac{\sigma_{11} + \sigma_{22} + \sigma_{33}}{3} \end{aligned} \tag{2.5}$$

When the isotropic material is considered, the plastic phenomenon can be described by the three invariants, which are independent of directions, of the deviatoric stress S_{ij}. These invariants J_1, J_2, and J_3 are defined as

$$\begin{aligned} J_1 &= S_{11} + S_{22} + S_{33} \\ J_2 &= S_{11}S_{22} + S_{22}S_{33} + S_{33}S_{11} - S_{12}S_{21} - S_{23}S_{32} - S_{31}S_{13} \\ J_3 &= \det(S_{ij}) \end{aligned} \tag{2.6}$$

Among the above invariants, the second invariant J_2 plays an important role in the flow theory. Meanwhile, the J_2 can be rearranged by stress tensor in the following form:

$$J_2 = \frac{1}{6}\left\{(\sigma_{11} - \sigma_{22})^2 + (\sigma_{22} - \sigma_{33})^2 + (\sigma_{33} - \sigma_{11})^2 + 6(\sigma_{12}^2 + \sigma_{23}^2 + \sigma_{31}^2)\right\} \quad (2.7)$$

2.2.2 Yield Condition

Various yield conditions have been proposed for different materials. For example, Mises and Tresca yield conditions are well known. In the case of J_2 flow theory, Mises yield condition is employed, and the equivalent stress is defined as

$$\bar{\sigma} = (3J_2)^{\frac{1}{2}} \quad (2.8)$$

So material will yield when the following condition is satisfied.

$$\bar{\sigma} = (3J_2)^{\frac{1}{2}} = \sigma_{yield} \quad (2.9)$$

where σ_{yield} is the current (after arbitrary plastic deformation) yield stress of the material.

In other words, material yields when the equivalent stress $\bar{\sigma}$ reaches the yielding stress, which is given by the one-dimensional stress-strain curve.

2.2.3 Strain Hardening Rule

Strain hardening or work hardening is a process by which the material grows stronger as it is deformed. In other words, yield stress after plastic deformation increases with the plastic strain. For a strain hardening material, the size and shape of the yield surface depend on the total history of deformation. Two approaches to describe the way that a material yields: isotropic hardening and kinematic hardening.

For isotropic hardening, the yield surface expands during plastic flow and this expansion is uniform in all directions about the origin in stress space. Thus initial shape and orientation is maintained. For kinematic hardening, the yield surface retains its original size, shape, and orientation with respect to the origin of the stress space, but the yield surface is assumed to undergo translation in the stress space. Kinematic hardening theory takes into account the Bauschinger effect and considers the material as a nonisotropic continuum. In reality, the hardening process often involves simultaneous translation and expansion of the yield surface, combining the isotropic hardening and nonlinear kinematic hardening approaches.

2.3 FAST COMPUTATION APPROACH

2.3.1 Iterative Substructure Method

In physical behavior, a local high-intensity power source caused by moving welding arc is applied to the steel plates during the welding process and then generates a sharp thermal profile in the nearby region and results in nonuniform distribution of plastic strains.

In the conventional FEM analysis, it is normal to consider the welding process as a nonlinear problem in the whole considered region and solve it for the entire process and total mesh domain. Actually, elastic or weakly nonlinear analysis can be carried out for the region away from the welding arc when the current temperature is lower than input temperature parameter, and the region with strongly nonlinear analysis moves together with the welding arc. Thus, an iterative substructure method (ISM) as shown in Figure 2.1 was developed as: a whole computational domain is divided into a weakly nonlinear region, a strongly nonlinear region, and boundary region, which are all changing with the moving welding heat source [25–27]. Meanwhile, the stiffness matrix of the weakly nonlinear region is updated only when it is necessary to economize the computing time. Wang et al. [66] employed the thermal elastic-plastic FE computation with ISM to predict the twisting welding deformation during the fabrication of

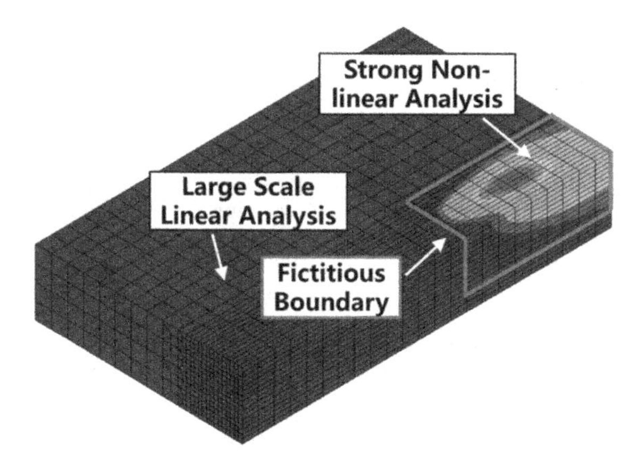

FIGURE 2.1 Image for iterative substructure method (ISM)

stiffened welded structure, and good agreement between computed results and measurement was observed.

2.3.2 Parallel Computation

For the conventional TEP FE analysis, the program is programmed and compiled with single-thread mode, which cannot utilize the advantage of multithreads of high-performance server to enhance the computational efficiency. Parallel computation with Open Multi-Processing technology was developed as a guided compilation processing scheme, which can be implemented for multithreads CPU with Shared Memory Operation System. Then, effective computation, which was already accepted and employed for computational mechanics, can be achieved by means of utilizing multithreads as shown in Figure 2.2.

With the OpenMP technology, parallel computation is becoming much easier for researchers and scientists, and the program can be supported with lots of computer languages such as C, C++, or FORTRAN. With two ranked matrices multiplication given in Eq. (2.10) as an example, parallel computation will transfer two ranked matrices to vectors, and then vector multiplication is carried out simultaneously with different threads as demonstrated in Eq. (2.11).

$$\begin{bmatrix} a_{11} & a_{12} \\ a_{21} & a_{22} \end{bmatrix} \begin{bmatrix} b_{11} & b_{12} \\ b_{21} & b_{22} \end{bmatrix} = \begin{bmatrix} c_{11} & c_{12} \\ c_{21} & c_{22} \end{bmatrix} \tag{2.10}$$

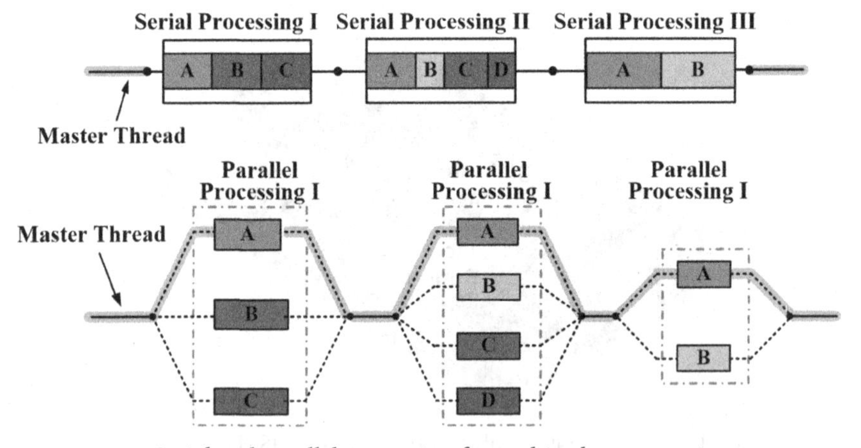

FIGURE 2.2 Serial and parallel processing for multitask computation

$$c_{11} = \begin{bmatrix} a_{11} \\ a_{12} \end{bmatrix}^T \begin{bmatrix} b_{11} \\ b_{21} \end{bmatrix} \quad c_{12} = \begin{bmatrix} a_{11} \\ a_{12} \end{bmatrix}^T \begin{bmatrix} b_{12} \\ b_{22} \end{bmatrix}$$
$$c_{21} = \begin{bmatrix} a_{21} \\ a_{22} \end{bmatrix}^T \begin{bmatrix} b_{11} \\ b_{21} \end{bmatrix} \quad c_{22} = \begin{bmatrix} a_{21} \\ a_{22} \end{bmatrix}^T \begin{bmatrix} b_{12} \\ b_{22} \end{bmatrix} \tag{2.11}$$

For the computational welding mechanics, the reading and writing process are normally completed with single thread mode, and operation of large range matrix is solved by OpenMP technology with multithreads mode. In detail, heat flux matrix during thermal analysis and mechanical stiffness matrix during mechanical analysis will be divided into submatrices and arranged for different threads for parallel computation. Thus, many more threads were employed for one computation, and computing time can be significantly reduced.

2.4 WELDING INHERENT STRAIN THEORY

Based on thermal elastic plastic FE analysis and experimental observations, Ueda et al [9, 15, 34, 35] concluded that welding residual stress and welding distortion are produced by the inherent strain ε^*. The inherent strain mostly depends on the joint parameters, such as type of welded joint, material properties, plate thickness, and heat input. For a specific welded joint, the inherent strain at each location is determined by the maximum temperature reached at this location during the welding process and the constraint provided by the surrounding material.

The total strain ε^{total} during the heating and cooling cycle of the welding process can be divided into the strain components given by Eq. (2.12), namely, elastic strain $\varepsilon^{elastic}$, plastic strain $\varepsilon^{plastic}$, thermal strain $\varepsilon^{thermal}$, creep strain ε^{creep}, and that produced through phase transformation ε^{phase}, respectively.

$$\varepsilon^{total} = \varepsilon^{elastic} + \varepsilon^{thermal} + \varepsilon^{plastic} + \varepsilon^{phase} + \varepsilon^{creep} \tag{2.12}$$

The total strain can be rearranged as a summation of the elastic strain and the inherent strain $\varepsilon^{inherent}$ which includes all the strain components except the elastic strain. In other words, the inherent strain $\varepsilon^{inherent}$ is defined as a summation of plastic strain, thermal strain, creep strain, and that caused by the phase transformation as given by Eq. (2.13). Especially for welded joints made of carbon steel, the inherent strain can be represented by the plastic strain because the strain induced by creep and solid-state phase transformation is much smaller.

$$\varepsilon - \varepsilon^{\text{elastic}} = \varepsilon^{\text{thermal}} + \varepsilon^{\text{plastic}} + \varepsilon^{\text{phase}} + \varepsilon^{\text{creep}} = \varepsilon^{\text{inherent}} = \varepsilon^{*} \qquad (2.13)$$

The distributions of inherent strain remain in a narrow area near the welding line. It may be seen that the longitudinal strain ε_x^* exists throughout the length of the welded joint and is constant in the middle part, where the effect of free ends may disappear; however, the transverse strain ε_y^* exists in relatively narrower region with much larger magnitude throughout the length of welded joint [40]. Figure 2.3 shows typical distributions of longitudinal and transverse plastic strain (inherent strain) on one cross-section normal to the welding line.

To obtain an idealized distribution and magnitude of inherent strain on a cross-section perpendicular to the welding line, theoretical analysis is employed.

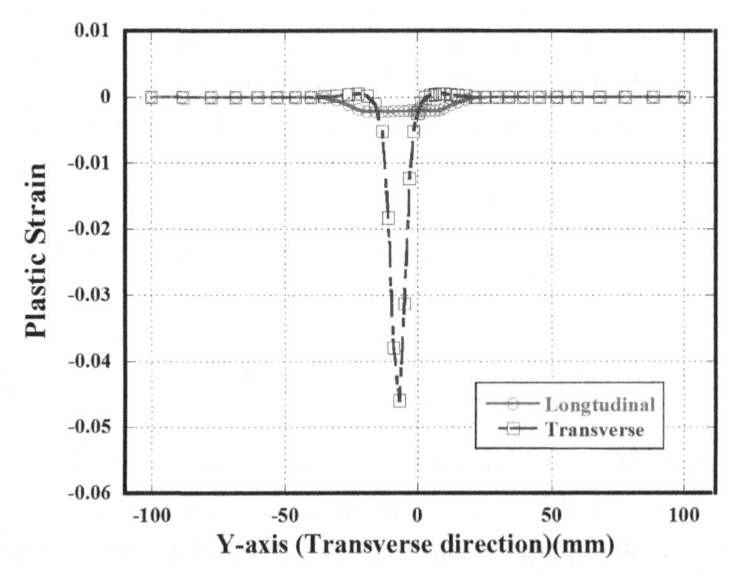

FIGURE 2.3 Distribution of plastic strain at the neutral axis of the middle transverse cross-section

Focusing on heat conduction in thin plate structures welded using high-speed welding, heat conduction can be approximated as one-dimensional (transverse direction of welded joint: y direction) heat conduction problem in solid, without heat loss by convection and radiation. The governing equation of heat conduction is expressed as Eq. (2.14) and the theoretical solution is given by Eq. (2.15) using Fourier transformation when the heated region is much narrower than the width of the whole plate.

$$a\frac{\partial^2 T(y,t)}{\partial y^2} = \frac{\partial T(y,t)}{\partial t} \quad (a = \frac{\lambda}{\rho c}) \tag{2.14}$$

$$T(y, t) = \frac{\eta Q}{\rho c\sqrt{4\pi a t}} e^{-\frac{y^2}{4at}} \tag{2.15}$$

where T(y, t) is the temperature which is a function of both position y and time t; λ, ρ, and c are the heat conductivity, the density, and the specific heat, respectively.

Differentiating the transient temperature with respect to the time t, as in Eq. (2.16), a point at a distance y from the welding line reaches a maximum temperature when the derivative ∂T(y,t)/∂t equals zero. The time at point y when the highest temperature occurs can be calculated, and this maximum temperature can be obtained as in Eq. (2.17) and Eq. (2.18).

$$\frac{\partial T(y,t)}{\partial t} = \frac{Q_{net}}{\rho c\sqrt{4\pi a}} \frac{\partial}{\partial t}(\frac{1}{\sqrt{t}} e^{-\frac{y^2}{4at}}) = \frac{Q_{net}}{\rho c\sqrt{4\pi a t}} e^{-\frac{y^2}{4at}}(\frac{y^2}{4at^2} - \frac{1}{2t}) \tag{2.16}$$

$$\frac{y^2}{4at^2} - \frac{1}{2t} = 0 \quad \text{when, } t = \frac{y^2}{2a} \tag{2.17}$$

$$T_{max}(y)\Big|_{t=\frac{y^2}{2a}} = \frac{Q_{net}}{\rho c\sqrt{2\pi e}} \frac{1}{y} \tag{2.18}$$

Noting that the welding region is constrained in the welding direction by the surrounding base metal, a simple mechanical model is proposed to investigate the mechanism which forms the inherent strain during the welding process.

When the welded joint is completely constrained in the welding direction and remains elastic during the heating stage of the thermal cycle, the relation between the stress σ and temperature T can be expressed by Eq. (2.19). The yield temperature TY can be expressed by Eq. (2.20). Furthermore, the heating temperature which is necessary to attain the yield stress after the cooling process is denoted T2 and is given by Eq. (2.21).

$$\sigma = -\alpha TE \tag{2.19}$$

$$T_Y = \frac{\sigma_Y}{\alpha E} = T_1 \tag{2.20}$$

$$T_2 = 2T_Y \tag{2.21}$$

where α is the coefficient of linear expansion and E and σ_Y are Young's modulus and the yield stress, respectively.

Considering these equations, the inherent strain in the welding direction can be derived for the three regions according to the maximum temperature T_{max} as given by Eq. (2.22), and corresponding longitudinal inherent strains ε_x^* are also obtained.

$$\begin{array}{cc} T_{max} \leq T_Y & \varepsilon_x^* = 0 \\ T_Y > T_{max} < 2T_Y & \varepsilon_x^* = -\alpha(T_{max} - T_Y) \\ 2T_Y \leq T_{max} & \varepsilon_x^* = -\alpha T_Y \end{array} \tag{2.22}$$

This model assumes fully restrained edges normal to the weld line. Also it does not take account of the temperature dependency of material properties. However, it is very useful to obtain a quick idealized solution. To take account of actual conditions, the inherent strain can be experimentally measured or evaluated using thermal elastic plastic FE analysis.

If the inherent strain is known, deformation and residual stress after the end of the welding process can be predicted by an elastic FE analysis using the inherent strain as an initial strain.

2.5 WELDING INHERENT DEFORMATION THEORY

As just mentioned above, the residual stress and the distortion after welding can be predicted using elastic FE analysis if the distribution of the inherent strain is known. To do this, a fine enough FE mesh, which is not desirable from practical point of view, is required. To avoid such a fine mesh, inherent deformation is proposed [34–36]. Noting that the displacement or the deformation is the integration of strain, inherent deformation, which is an integration of the inherent strain, can be used to predict welding distortion without a significant loss of accuracy.

Similar to the inherent strain, the inherent deformation mostly depends on the joint parameters. The influence of the length and the width of a welded joint is small if the size of the plate is large enough [82, 83]. When the edges effect is ignored, the components of inherent deformation can be approximated as constant values along the welding line. These constant values are introduced into the elastic model as loads (forces and displacements) to predict welding distortion. Computed results are in good agreement with experimental measurements.

The inherent deformation can be evaluated as the integration of the longitudinal inherent strain ε_x^* in the welding direction and the transverse inherent strain ε_y^* in the transverse direction distributed on the cross-section normal to the welding line according to the following equations.

$$
\begin{aligned}
\delta_x^* &= \frac{1}{h}\iint \varepsilon_x^* \mathrm{d}y\mathrm{d}z \\
\delta_y^* &= \frac{1}{h}\iint \varepsilon_y^* \mathrm{d}y\mathrm{d}z \\
\theta_x^* &= \frac{12}{h^3}\iint (z-\frac{h}{2})\varepsilon_x^* \mathrm{d}y\mathrm{d}z \\
\theta_y^* &= \frac{12}{h^3}\iint (z-\frac{h}{2})\varepsilon_y^* \mathrm{d}y\mathrm{d}z
\end{aligned}
\tag{2.23}
$$

where δ_x^* and δ_y^* are the inherent deformation in the longitudinal and the transverse directions, respectively; θ_x^* and θ_y^* are the inherent bending deformation in the longitudinal and the transverse directions, respectively; h is the thickness of the welded joint; and x, y, z are the welding direction, transverse direction, and thickness direction, respectively.

The longitudinal inherent shrinkage has a different nature from the other inherent deformation components. This is caused by the constraint provided by the surrounding material that prevents free shrinkage along the weld line. Therefore, the longitudinal inherent deformation appears as a tensile force acting in the weld line that is referred to as the Tendon Force. Other components are subjected to only a small constraint and appear as deformation.

According to the common understanding in welding mechanics, longitudinal shrinkage and buckling distortion due to welding are produced by the longitudinal inherent shrinkage, more specifically, the longitudinal inherent shrinkage force [65]. This force has a strong tensile nature and small acting area resembling a tendon. This is why it is referred to as the tendon force.

Compressive stresses are developed in the area of the plate other than the area around the weld bead to achieve equilibrium with the Tendon Force. These compressive stresses are responsible for buckling.

When the material properties of carbon steel at room temperature are used, the tendon force can be expressed as given by Eq. (2.24). This equation has an approximately similar coefficient as that in the empirical formula given by Eq. (2.25).

$$F_{tendon} = -0.335\frac{E\alpha}{\rho c}\eta Q = -0.235\eta Q = -0.235 Q_{net} \tag{2.24}$$

where η is the welding heat efficiency;
E is Young's modulus, $E = 2.1\times10^{6}\,\text{MPa}$;
α is the coefficient of line expansion, $\alpha = 1.2\times10^{-5}\,/\,^{\circ}\text{C}$;
ρ is the density, $\rho = 7800\,\text{kg}/\text{m}^{3}$;
and c is the specific heat of material, $c = 4.6\times10^{2}\,\text{J}/\text{kg}\cdot{}^{\circ}\text{C}$.

Furthermore, the relation between the longitudinal inherent deformation δ_L^* and the tendon force F_{tendon} (longitudinal inherent shrinkage force) can be derived from their definitions given by Eq. (2.25).

$$F_{tendon} = \int E\times\varepsilon_L^* dA = E\times h\times\frac{1}{h}\int\varepsilon_L^* dA = Eh\,\delta_L^* \tag{2.25}$$

where ε_L^* and $\delta_L^* = \frac{1}{h}\int\varepsilon_L^* dA$ are the inherent strain and inherent deformation in the longitudinal direction, respectively, and h is the thickness of the welded joint.

2.6 ELASTIC STABILITY OF WELDING PLATE

Occurrence of welding-induced buckling is a kind of nonlinear response, which is also considered as a stability problem. This phenomenon can be predicted and represented using the large deformation theory. The equation relating to the strain and displacement is essential to describe this nonlinear response. If the small deformation is assumed, the strains are given as a linear function of displacements. In the case of the large deformation theory, the nonlinear relation between strain and displacement will be considered, and Green-Lagrange strain which is the second-order function of the displacements should be used as given by Eq. (2.26). From the expressions of these strains, the first-order term represents the linear response, and the second-order term is essential for the nonlinear phenomenon under large deformation.

$$
\begin{aligned}
\varepsilon_x &= \frac{\partial u}{\partial x} + \frac{1}{2}\left\{\left(\frac{\partial u}{\partial x}\right)^2 + \left(\frac{\partial v}{\partial x}\right)^2 + \left(\frac{\partial w}{\partial x}\right)^2\right\} \\
\varepsilon_y &= \frac{\partial v}{\partial y} + \frac{1}{2}\left\{\left(\frac{\partial u}{\partial y}\right)^2 + \left(\frac{\partial v}{\partial y}\right)^2 + \left(\frac{\partial w}{\partial y}\right)^2\right\} \\
\varepsilon_z &= \frac{\partial w}{\partial z} + \frac{1}{2}\left\{\left(\frac{\partial u}{\partial z}\right)^2 + \left(\frac{\partial v}{\partial z}\right)^2 + \left(\frac{\partial w}{\partial z}\right)^2\right\} \\
\gamma_{xy} &= \frac{\partial u}{\partial y} + \frac{\partial v}{\partial x} + \left\{\left(\frac{\partial u}{\partial x}\right)\left(\frac{\partial u}{\partial y}\right) + \left(\frac{\partial v}{\partial x}\right)\left(\frac{\partial v}{\partial y}\right) + \left(\frac{\partial w}{\partial x}\right)\left(\frac{\partial w}{\partial y}\right)\right\} \\
\gamma_{yz} &= \frac{\partial v}{\partial z} + \frac{\partial w}{\partial y} + \left\{\left(\frac{\partial u}{\partial y}\right)\left(\frac{\partial u}{\partial z}\right) + \left(\frac{\partial v}{\partial y}\right)\left(\frac{\partial v}{\partial z}\right) + \left(\frac{\partial w}{\partial y}\right)\left(\frac{\partial w}{\partial z}\right)\right\} \\
\gamma_{zx} &= \frac{\partial w}{\partial x} + \frac{\partial u}{\partial z} + \left\{\left(\frac{\partial u}{\partial z}\right)\left(\frac{\partial u}{\partial x}\right) + \left(\frac{\partial v}{\partial z}\right)\left(\frac{\partial v}{\partial x}\right) + \left(\frac{\partial w}{\partial z}\right)\left(\frac{\partial w}{\partial x}\right)\right\}
\end{aligned}
\tag{2.26}
$$

where ε_x, ε_y, and ε_z are Green-Lagrange strain in x, y, and z directions, respectively; γ_{xy}, γ_{yz}, and γ_{zx} are the shear strain on the x-y, y-z, and z-x planes, respectively; u, v, and w are the displacement in x, y, and z directions, respectively.

It is well known that eigenvalue analysis is always employed to predict the buckling force of a structure, which is assumed to be an ideal linear elastic body. In the classical eigenvalue analysis, eigenvalues are computed with regard to the applied compressive force and constraints of a given system. For a basic structural configuration, each applied combination of forces has a minimum critical buckling value at which the structure buckles, and a corresponding buckling mode. Eigenvalue analysis also gives other higher-order buckling modes and the associated critical buckling values. Theoretically, there are values and modes whose numbers are equal to the numbers of degree of freedom in the considered system.

The basic variables, such as the displacement $u_{t+\Delta t}$, the strain $\varepsilon_{t+\Delta t}$, and the stress $\sigma_{t+\Delta t}$ at time $t+\Delta t$ can be decomposed into the sum of their values at time t and their increments, i.e.,

$$\begin{aligned} u_{t+\Delta t} &= u_t + \Delta u \\ \varepsilon_{t+\Delta t} &= \varepsilon_t + \Delta\varepsilon = \varepsilon_t + \Delta^1\varepsilon + \Delta^2\varepsilon \\ \sigma_{t+\Delta t} &= D\varepsilon_{t+\Delta t} = \sigma_t + D\Delta\varepsilon = \sigma_t + D\Delta^1\varepsilon + D\Delta^2\varepsilon \end{aligned} \tag{2.27}$$

where D is the elastic matrix (stress-strain matrix) and $\Delta^1\varepsilon$ and $\Delta^2\varepsilon$ are the first-order (linear) and the second-order (nonlinear) terms of strain increment, respectively.

When Eq. (2.26) is used, and taking ε_x for an example, it is shown that

$$\begin{aligned} \varepsilon_x(t+\Delta t) &= \frac{\partial u_{t+\Delta t}}{\partial x} + \frac{1}{2}\left(\frac{\partial u_{t+\Delta t}}{\partial x}\right)^2 + \frac{1}{2}\left(\frac{\partial v_{t+\Delta t}}{\partial x}\right)^2 + \frac{1}{2}\left(\frac{\partial w_{t+\Delta t}}{\partial x}\right)^2 \\ &= \varepsilon_x(t) + \Delta^1\varepsilon_x + \Delta^2\varepsilon_x \end{aligned} \tag{2.28}$$

where

$$\begin{aligned} \varepsilon_x(t) &= \frac{\partial u_t}{\partial x} + \frac{1}{2}\left(\frac{\partial u_t}{\partial x}\right)^2 + \frac{1}{2}\left(\frac{\partial v_t}{\partial x}\right)^2 + \frac{1}{2}\left(\frac{\partial w_t}{\partial x}\right)^2 \\ \Delta^1\varepsilon_x &= \frac{\partial \Delta u}{\partial x} + \frac{\partial u_t}{\partial x}\frac{\partial \Delta u}{\partial x} + \frac{\partial v_t}{\partial x}\frac{\partial \Delta v}{\partial x} + \frac{\partial w_t}{\partial x}\frac{\partial \Delta w}{\partial x} \\ \Delta^2\varepsilon_x &= \frac{1}{2}\left(\frac{\partial \Delta u}{\partial x}\right)^2 + \frac{1}{2}\left(\frac{\partial \Delta v}{\partial x}\right)^2 + \frac{1}{2}\left(\frac{\partial \Delta w}{\partial x}\right)^2 \end{aligned}$$

To derive the governing equation for eigenvalue analysis, the minimum potential energy theorem is employed. The total energy of the system at times t and t + Δt are given by Eqs. (2.29) and (2.30), respectively.

$$\pi(u_t) = \int \frac{1}{2} \varepsilon_t^T D \varepsilon_t dv - \int f_t u_t ds \tag{2.29}$$

$$\begin{aligned} \pi(u_t + \Delta u) = & \int \frac{1}{2} (\varepsilon_t + \Delta^1\varepsilon + \Delta^2\varepsilon)^T D (\varepsilon_t + \Delta^1\varepsilon + \Delta^2\varepsilon) dv \\ & - \int (f_t + \Delta f)(u_t + \Delta u) ds \\ & = \pi(u_t) + \Delta\pi(\Delta u) \end{aligned} \tag{2.30}$$

Since the potential energy at time t is already known and fixed, the condition for the potential energy $\pi(u_t + \Delta t)$ to be a minimum is equivalent to that condition for its increment $\Delta\pi(\Delta u)$, given by Eq. (2.31), in which the higher-order terms are neglected. Eq. (2.31) can be rewritten in matrix form as shown in Eq. (2.32).

$$\begin{aligned} \Delta\pi(\Delta u) &= \pi(u_t + \Delta t) - \pi(u_t) \\ &= \int \frac{1}{2} \{(\Delta^1\varepsilon)^T D(\Delta^1\varepsilon) + 2D\varepsilon_t \Delta^1\varepsilon + 2D\varepsilon_t \Delta^2\varepsilon\} dv - \int (f_t \Delta u + \Delta f u_t) ds \\ &= \int \frac{1}{2} \{(\Delta^1\varepsilon)^T D(\Delta^1\varepsilon) + 2\sigma_t \Delta^2\varepsilon\} dv - \int \Delta f u_t ds + \int \sigma_t \Delta^1\varepsilon dv \\ &\quad - \int f_t \Delta u ds \end{aligned} \tag{2.31}$$

$$\begin{aligned} \Delta\pi(\Delta u) = & \frac{1}{2}\{\Delta u\}^T [K_1(u_t)] \{\Delta u\} + \frac{1}{2}\{\Delta u\}^T [K_2(\sigma_t)] \{\Delta u\} \\ & - \{\Delta f\}^T \{\Delta u\} + \{F\}^T \{\Delta u\} - \{f_t\}^T \{\Delta u\} \end{aligned} \tag{2.32}$$

Where

$$\frac{1}{2}\{\Delta u\}^T[K_1(u_t)]\{\Delta u\} = \int \frac{1}{2}(\Delta^1\varepsilon)^T D(\Delta^1\varepsilon)dv$$
$$\frac{1}{2}\{\Delta u\}^T[K_2(\sigma_t)]\{\Delta u\} = \int D\varepsilon_t \Delta^2\varepsilon dv = \int \sigma_t \Delta^2\varepsilon dv$$
$$\{F\}^T\{\Delta u\} = \int D\varepsilon_t \Delta^1\varepsilon dv$$
$$\{\Delta f\}^T\{\Delta u\} = \int \Delta f \Delta u_t dv$$
$$\{f_t\}^T\{\Delta u\} = \int f_t \Delta u dv$$

From the condition for the minimum value of $\Delta\pi(\Delta u)$, the following equation is derived.

$$\frac{\partial \Delta\pi(\Delta u)}{\partial \Delta u} = [K_1(u_t)]\{\Delta u\} + [K_2(\sigma_t)]\{\Delta u\} + \{F\} - \{f_t\} - \{\Delta f\} = 0 \tag{2.33}$$

Due to the equilibrium of the system at time t:

$$\{F\} - \{f_t\} = 0 \tag{2.34}$$

Because of the fact that the buckling occurs without increase of external forces, Eq. (18) is satisfied.

$$\{\Delta f\} = 0 \tag{2.35}$$

If the system buckles under an internal force $\lambda\{\sigma_t(\text{applied})\}$, Eq. (2.33) is reduced to an eigenvalue problem given by Eq. (2.36).

$$[K_1(u_t)]\{\Delta u\} + \lambda[K_2(\lambda\sigma_t(\text{applied}))]\{\Delta u\} = 0 \tag{2.36}$$

From Eq. (19), it is clear that buckling is an eigenvalue problem. In case of welding, the stress σ_t(applied) is the stress produced by the inherent deformation associated with the welding and the parameter λ is the eigenvalue to be determined. Eq. (2.37) gives the definition of eigenvalue parameter λ. When the critical stress becomes equal $\lambda\sigma_t$(applied), the structure buckles. This means that the structure has already buckled when λ is less than 1.0, or critical stress is less than applied stress.

$$\lambda = \frac{\sigma(\text{critical})}{\sigma(\text{applied})} \tag{2.37}$$

2.7 INDUCTION HEATING TECHNIQUE

In nature, some metals such as iron and steel have electromagnetic characteristics, and induction heat will be generated as physical response when high-frequency oscillatory current appears nearby due to their best magnetic hysteresis [137]. Generally, hysteresis heating occurs when the temperature is lower than Curie temperature, and magnetic materials will retain their electric-magnetic characteristic and induction heating behavior.

This physical mechanism can be clarified with high permeability, which is beneficial for induction heating in magnetic materials and appears when

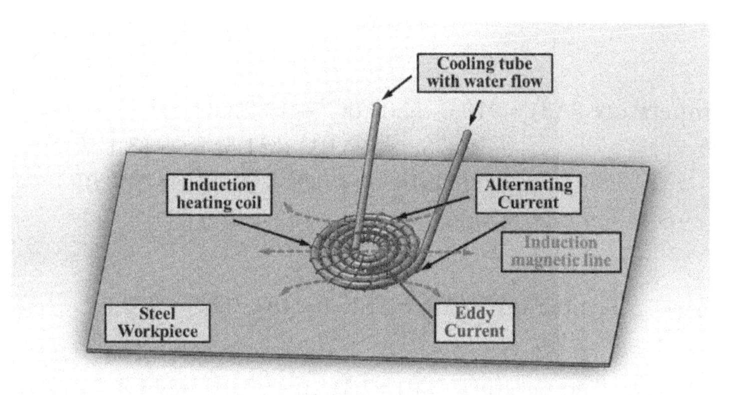

FIGURE 2.4 Image of physical behavior of induction heating

temperature is lower than Curie temperature. Therefore, induction heating is a physical process of heat generation and conduction on magnetic metals with electromagnetic induction behavior [137]. In detail, induction heat is generated inside of steel workpiece itself due to the eddy current, which is the result of high-frequency oscillatory current of induction heating coil and rapidly alternating magnetic field.

Then, steel workpiece can be heated without any external heat source. The heat amount generated by induction heating is significantly influenced by the distance between induction heating coil and steel workpiece. Meanwhile, heat losses will occur due to the physical behaviors of thermal convection and radiation. In addition, the frequency of oscillatory current dominantly determines the penetrated depth, as well as the region of induction eddy current that exists in the thickness direction of heated steel workpiece.

2.8 SOLID PHASE TRANSFORMATION DURING WELDING

Generally, welded joint can be divided into three portions: welded zone, heat affected zone (HAZ), and base material zone. HAZ can be defined as the portion of base material which was not melted during welding but whose microstructure and mechanical properties were altered by the welding heating. Based on the chemical compositions of examined material, the critical temperatures for solid phase transformation, as well as microstructure evolution, can be calculated as follows:

$$\begin{aligned}\text{Temperature (A1)} = {} & 723.0 - 10.7\text{Mn} - 16.9\text{Ni} + 29.0\text{Si} \\ & + 16.9\text{Cr} + 290.0\text{As} + 6.4\text{W}\end{aligned} \tag{2.38}$$

$$\begin{aligned}\text{Temperature (A3)} = {} & 910.0 - 203.0\text{C}^2 - 15.2\text{Ni} \\ & + 44.7\text{Si} + 104.0\text{V} + 31.5\text{Mo} + 13.1\text{W} \\ & - 30.0\text{Mn} - 11.0\text{Cr} - 20.0\text{Cu} + 700.0\text{P} \\ & + 400.0\text{Al} + 120.0\text{As} + 400.0\text{Ti}\end{aligned} \tag{2.39}$$

$$\text{Temperature (TS)} = 1527.0 - 181.356\text{C} - 273.0 \tag{2.40}$$

$$\text{Temperature (TP)} = 1810.0 - 90.0\text{C} - 273.0 \tag{2.41}$$

$$\text{Temperature (BS)} = 656.0 - (58.0\text{C} + 35.0\text{Mn}) - (75.0\text{Si} + 15.0\text{Ni}) - (34.0\text{Cr} + 41.0\text{Mo}) \quad (2.42)$$

$$\text{Temperature (MS)} = 561.0 - (474.0\text{C} + 35.0\text{Mn}) - 17.0\text{Ni} - (17.0\text{Cr} + 21.0\text{Mo}) \quad (2.43)$$

where the units are all degrees centigrade (°C), and the chemical symbols are their mass percentages. A1 and A3 mean the temperature for Pearlite and Ferrite transformations, TS means the temperature for precipitate dissolution, TP means melting point temperature, and BS and MS mean the temperature for Austenite decomposition to Bainite and Martensite.

Therefore, the computed thermal cycles of solid elements model can be employed to distinguish the HAZ during FE analysis; the points whose maximal temperature is higher than A1 point temperature (about 727°C) can be considered HAZ due to the occurrence of solid phase transformation and microstructure alternation. In detail, the computed welding thermal cycle was used to predict HAZ hardness based on the microstructure evolution method. According to the continuous cooling transformation diagram of the HAZ during cooling procedure, as displayed in Figure 2.5, the Austenite decomposition products during cooling process are dominantly determined by the cooling rate.

With the different cooling rates (v), as shown in curves A, B, C, and D, different Austenite transformation products formed. In detail, a fast cooling rate shown by curve A led to the formation of Martensite, and as the cooling rate reduced as shown by curves B, C, and D, other transformation products such as Ferrite, Pearlite, Bainite, and Martensite were formed. Therefore, the HAZ hardness could be calculated when the hardness and volume fraction of various transformation products above were known. The calculation formula was determined as Eq. (2.44).

$$V_{\text{decomposition}} = X_M \times V_M + X_B \times V_B + X_F \times V_F + X_P \times V_P + X_{RA} \times V_{RA} \quad (2.44)$$

where, H_M, H_B, and H_{AFP} represent the hardness of Martensite, Bainite, and Austenite-Ferrite-Pearlite, respectively; they are the functions of alloy element content and cooling rate, given as follows:

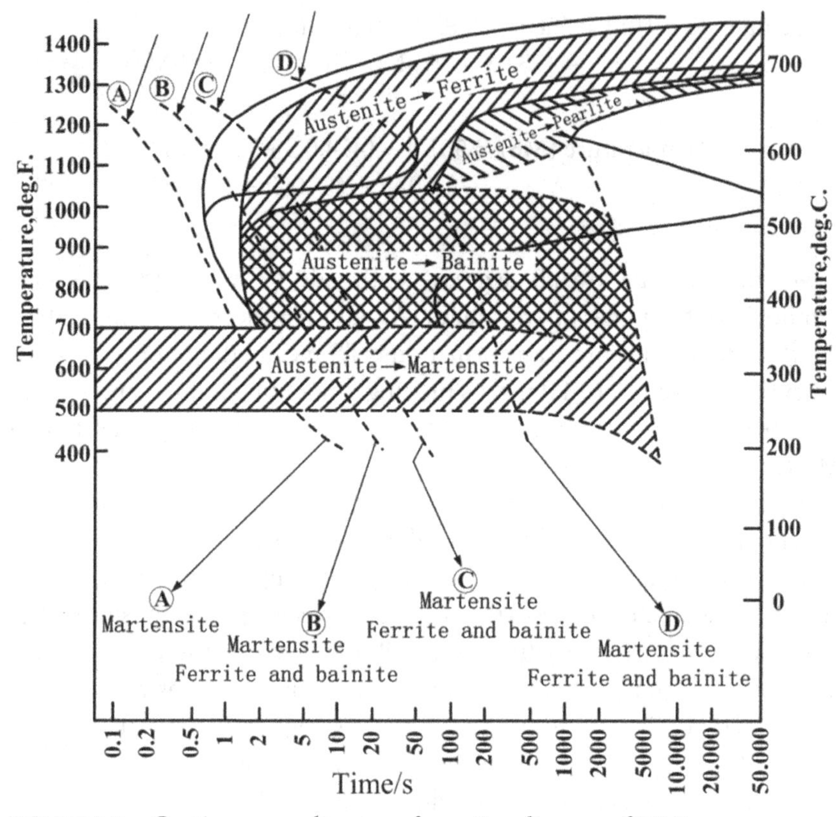

FIGURE 2.5 Continuous cooling transformation diagram of HAZ

$$H_M = 127 + 949C + 27Si + 11Mn + 8Ni + 16Cr + 21\log_{10} v \quad (2.45)$$

$$\begin{aligned} H_B = &- 323 + 185C + 330Si + 153Mn + 65Ni + Cr + 191Mo \\ &+ (89 + 53C - 55Si - 22Mn - 10Ni - 20Cr - 33Mo)\log_{10} v \end{aligned} \quad (2.46)$$

$$\begin{aligned} H_{AFP} = &42 + 223C + 53Si + 30Mn + 12.6Ni + 7Cr + 191Mo \\ &+ (10 - 19Si + 4Ni + 8Cr + 130V)\log_{10} v \end{aligned} \quad (2.47)$$

$$v = (800 - 500 / t_{8/5}) \, 1/3600,$$

$$t_{8/5} = q/2\pi\lambda v \, (1/500 - T_0 - 1/800 - T_0) \quad (2.48)$$

T_0 is the ambient or preheat temperature (°C); λ is the thermal conductivity ($J / m \cdot sec°C$); q/v is the heat input (J/m); and v is the cooling rate.

X_M, X_B, and X_{AFP} are the volume fractions of Martensite, Bainite, and Austenite-Ferrite-Pearlite. All of them are controlled by the following equation.

$$\frac{dx}{dt} = B_r(G,T)X^m(1-X)^n \tag{2.49}$$

where X is the volume fraction of the Austenite decomposition product; Br is an effective rate coefficient; G is the austenite grain size index number; T is the temperature in K; t is the time in seconds; and m and n are semiempirical coefficients.

Moreover, the Austenite decomposing products on cooling process were determined by their own formation start temperature. Specifically speaking, the upper critical temperature (A3) was the Ferrite start temperature and the lower critical temperature (A1) was the Pearlite start temperature; BS was the Bainite start temperature and MS was the Martensite start temperature. All of them were the functions of carbon and alloy element content. The critical tempcratures divided the welding thermal cycle into eight regions, as illustrated in Figure 2.6. The microstructure evolution in HAZ during one thermal cycle was composed of two sections: (1) Austenization on heating process and (2) Austenite decomposition during cooling process.

On the foundation of the microstructure evolution, the hardenability algorithm was modified to predict the HAZ hardness of butt-welded joint obtained by multipass welding. To improve computation efficiency, the welding thermal cycle with a maximum temperature higher than A1 will be taken into account. In addition, the HAZ microstructure is still a mixture of Austenite-Ferrite-Pearlite on cooling process when the cooling rate is zero.

For multipass welding, the retained Austenite-Ferrite-Pearlite would continue getting Austenization on heating process and transforming to its decomposing products during cooling process. Therefore, the total hardness will be given as Eq. (2.50):

$$H_{total} = \sum_{i=1}^{n} (H_{Mi} X_{Mi} + H_{Bi} X_{Bi}) + H_{AFP} X_{AFP} \tag{2.50}$$

where n is the number of considered welding thermal cycles.

As a whole, the examined point can be distinguished as HAZ when its maximal temperature is higher than A1 point temperature (about 727°C),

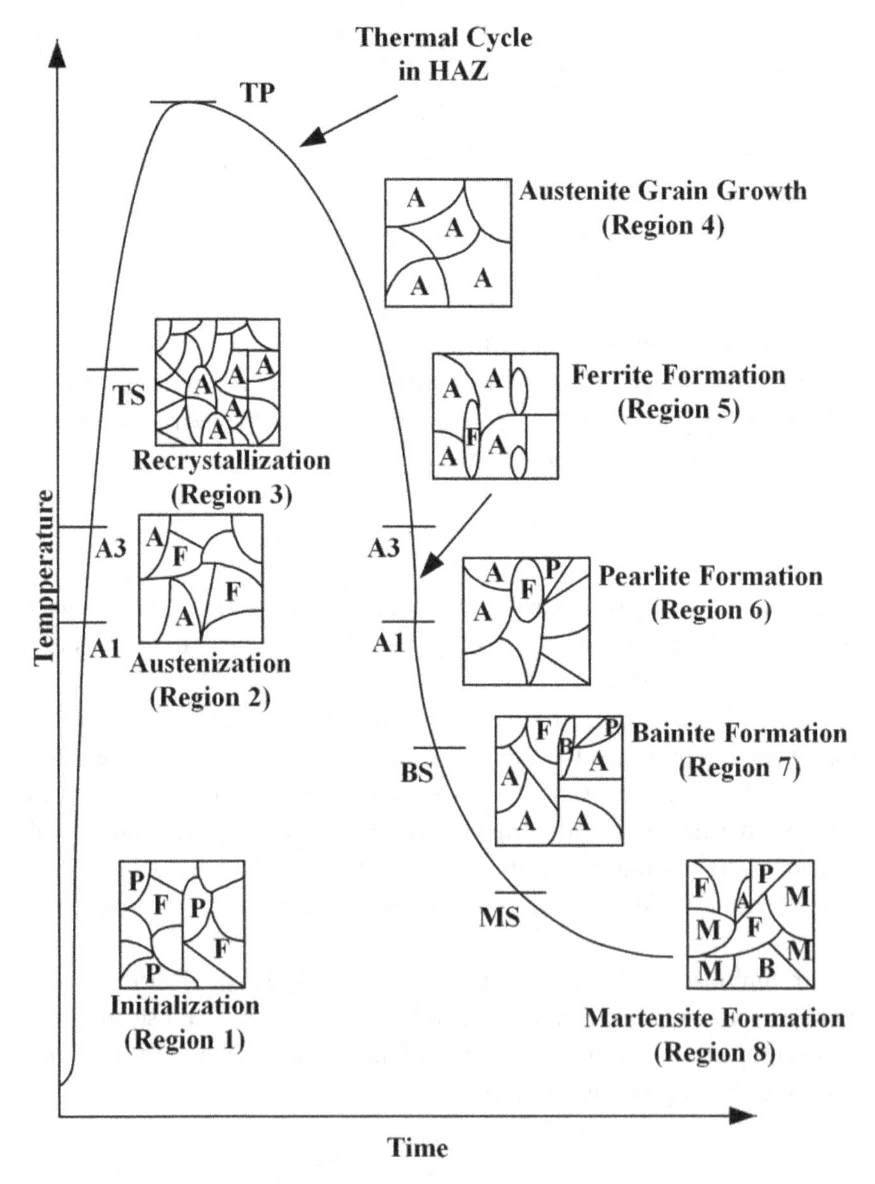

FIGURE 2.6 Microstructure evolution of HAZ during one thermal cycle

which will transform to Austenite due to solid phase transformation during heating. Then, the Austenite decomposition to Bainite, Martensite, Ferrite, or Pearlite will occur during cooling with the proposed microstructure evolution approach. Because of the different hardness and percentage of each solid phase or microstructure, the hardness of examined point can be eventually evaluated with Eq. (2.50).

2.9 SUMMARY

Elementary theories and methods for investigations of welding distortion and residual stress were systematically introduced, while transient thermal elastic plastic FE computation with advanced computation technique is presented to effectively examine the thermal and mechanical responses due to welding. Concept of welding inherent strain and inherent deformation was also presented to clarify the generation mechanism of welding-induced buckling, and induction heating as heating source of transient thermal tension technique was also introduced. To accurately evaluate the welding residual stress of high tensile strength steel, solid phase transformation as well as microstructure evolution was eventually presented.

CHAPTER 3

C-Mn Thin Plates Butt Welding and Measurement

In order to reduce the top weight with thin plates, high tensile strength steel such as C-Mn alloy steel with perfect mechanical performance is commonly employed in the fabrication of lightweight structure such as advanced ship, while welding method is used to assemble different pieces or plates as welded joints and structures. As the typical welded joint, butt and fillet welding are usually considered to examine the thermal and mechanical behaviors; in particular, welding distortion as well as out-of-plane welding buckling will be concentrated for the thin plate welded joints with high tensile strength steel. Thus, a typical high tensile strength steel of AH36 with yield stress of 360 MPa was selected to investigate the welding distortion and accurate fabrication of the abovementioned typical welded joints.

3.1 MATERIALS AND WELDING METHOD

C-Mn steel of AH36 as a typical kind grade of high tensile strength steel has a wide application for ship and offshore structures. Table 3.1 shows the chemical component of AH36, and its common physical properties such as yield stress, linear thermal expansion coefficiency, and ultimate strength are also given in Table 3.2.

DOI: 10.1201/9781003442523-3

TABLE 3.1 Chemical component of AH36.

C	Si	Mn	P	S	Cu	Cr	Ni	Nb
<0.18	<0.5	0.9–1.6	<0.03	<0.03	<0.35	<0.2	<0.4	0.02–0.05

TABLE 3.2 Mechanical performance of AH36.

Yield Stress	Ultimate Strength	Fracture Strength	Elongation
416 MPa	490 MPa	430 MPa	21%

In general, there are lots of welding methods such as gas metal arc welding (GMAW), flux cored arc welding (FCAW), submerged arc welding (SAW), Tungsten inert-gas arc welding (TIG), and gas tungsten arc welding (GTAW), to join the AH36 steel together with good mechanical performance of welded joint. With the advantages of flexible features and low cost, gas-shielded arc welding with CO_2 as shielding gas is usually employed in the actual fabrication of shipbuilding. Thus, gas-shielded arc welding with CO_2 will be used for the later welding experiment of AH36 thin plates.

3.2 INDUCTION HEATING EXPERIMENT AND COMPUTATION

Since penetration depth of induction heating on the plate thickness is mainly determined by current oscillation frequency of equipment, much higher oscillation frequency will generate much lower heated depth. The temperature difference between top and bottom plate surfaces as well as temperature gradient in the plate thickness direction will generate the out-of-plane bending deformation. Thus, transient thermal tension with additional induction heating should make the heating source roughly penetrate the plate thickness without generation of bending deformation. Induction heating and measurement system as shown in Figure 3.1 was established and employed to examine the thermal and mechanical responses.

In addition, induction heating coil was fixed on a moveable car with special chucking appliance, which can move with uniform speed of 0–25 mm/s on the track. Transient temperature and thermal cycle will be measured with K-type thermocouple of platinum-rhodium alloy, and then the measured data will be converted to a digital signal and transmitted to

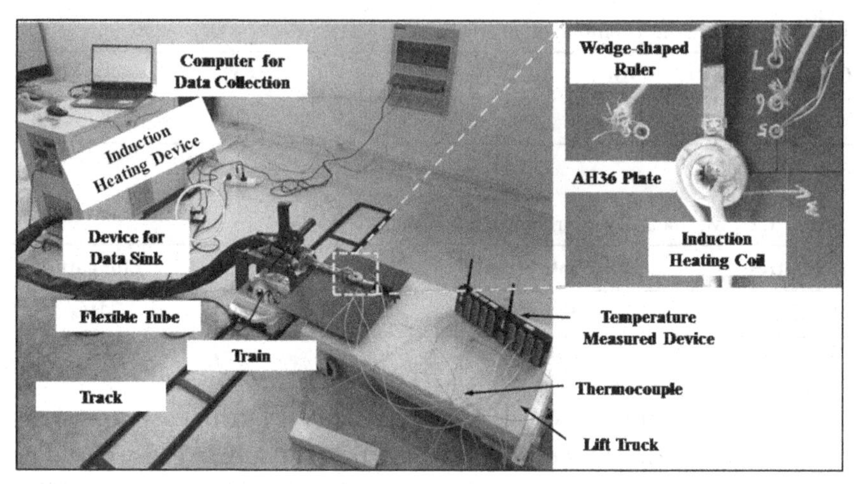

FIGURE 3.1 Induction heating and measurement system

the personal computer with time increment of 0.25 seconds. The metal material of induction heating is high tensile strength steel with grade of AH36, and the dimension size of steel plate is 400 mm in length, 300 mm in width, and 5 mm in thickness. Induction heating coil will move with uniform speed in the longitudinal direction of steel plate, while its diameter is about 52 mm. Moreover, distance between induction heating coil and steel plate is roughly fixed to be 8 mm during the entire induction heating experiment, which is achieved by means of height of lifter device and measurement of wedge ruler as demonstrated in Figure 3.1.

As the main component for induction heating, there is one induction heating equipment as given in Figure 3.2(a) with output power of 40 kW, while the medium frequency is about 1–20 kHz, penetrated depth is about 3–10 mm, and a flexible tube with length of 2 meters was also designed to enhance the operation space of induction heating coil. Moreover, the induction heating coil is designed as tube with copper, and in order to prevent the high temperature of coil during induction heating, cycle cooling water was employed in the copper tube, which is supported by a cooling machine as given in Figure 3.2(b) with compressive force of 20 HP to make sure the temperature range from 3°C to 30°C.

Induction heating equipment with medium frequency could control the output power by means of working current. Output power will increase with the increase of working current. When the moving speed of induction heating coil is constant, maximal temperature of steel plate will be determined by output power as well as working current.

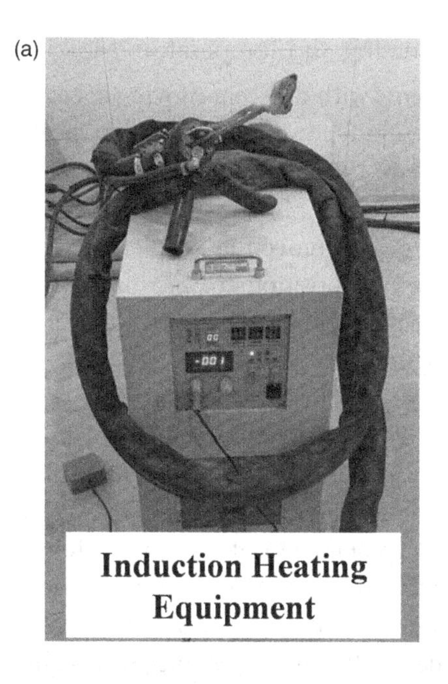

FIGURE 3.2 Induction heating system

3.2.1 Induction Heating and Temperature Measurement

A series of experiments with different moving speeds of induction heating coil and different output powers of induction heating equipment were carried out to measure the thermal cycle and evaluate temperature field distribution.

At first, the K-type thermocouple of platinum-rhodium alloy should be fixed at the designed position to measure the thermal cycle during induction heating, which will be actualized with tack welding machine as demonstrated in Figure 3.3.

As shown in Figure 3.4, the designed position to measure thermal cycle during induction heating can be observed. There are totally eight positions for temperature measurement with different transverse distances with respect to center line of induction heating. In addition, measured points with No. 1–3 (green color) are located at the bottom surface of steel plate, and measured points with No. 4–8 (red color) are located at the top surface of steel plate as demonstrated in Figure 3.4 by different colors. This kind of measured plan to design the thermocouple position mainly depends on the influence of movement of induction heating coil and ensures the experiment process of induction heating on steel plate.

When the moving speed of induction heating coil is fixed, the reached maximal temperature of heated region will be determined by output power of induction heating, as well as the distance between induction heating coil and steel plate together, which will significantly influence the thermal efficiency of induction heating.

In addition, plastic strain due to induction heating will be generated during the heating and cooling processes when the maximal temperature

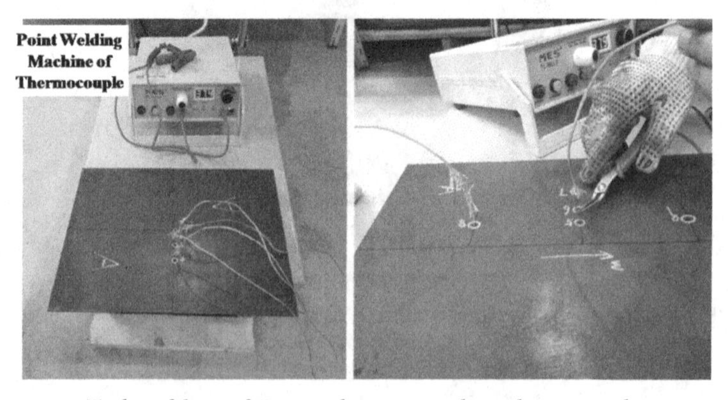

FIGURE 3.3 Tack welding of K-type thermocouple at the proper location

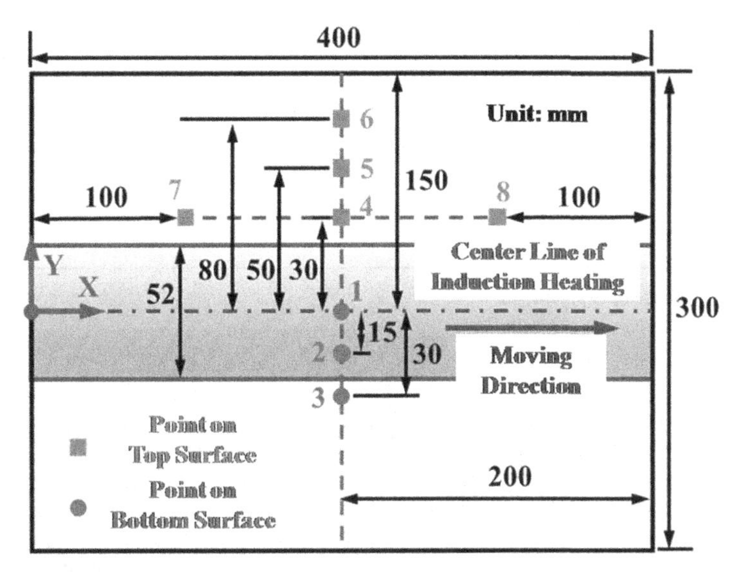

FIGURE 3.4 Distribution of thermocouples during induction heating

of examined region is higher than the yield temperature. Obviously, the output power of induction heating as well as the working current can be obtained from the induction heating equipment; the thermal efficiency will not be determined directly and net heat input due to induction heating will not be accurately calculated for the subsequential investigation.

Therefore, there is a series of experiments with different output power of induction heating, which was designed to determine the exact value of thermal efficiency by means of measurement of thermal cycles of some examined points with thermocouples. In order to avoid the measurement tolerance due to tack welding of thermocouples as well as constant distance between induction heating coil and steel plate, experiment of induction heating will be carried out as follows. One constant moving speed of induction heating will be employed only for one steel plate, and several output powers will be considered during actual experiment, while the information in detail is summarized in Table 3.3. There are four kinds of moving speeds of induction heating coil and five experimental currents instead of output power of induction heating.

Moreover, each experiment of induction heating with different output power and constant moving speed was carried out and the planeness of examined steel plate should be checked after cooling down to room temperature. There will be not plastic strain and out-of-plane distortion due

TABLE 3.3 Experimental parameters of induction heating.

Speed \ Current	400 A	450 A	550 A	650 A	780 A
3.0 mm/s	√	√	×	×	×
5.0 mm/s	×	√	√	√	×
7.8 mm/s	×	√	√	√	√
9.8 mm/s	×	√	√	√	√

√ *means existence of experiment.*
× *means no existence of experiment.*

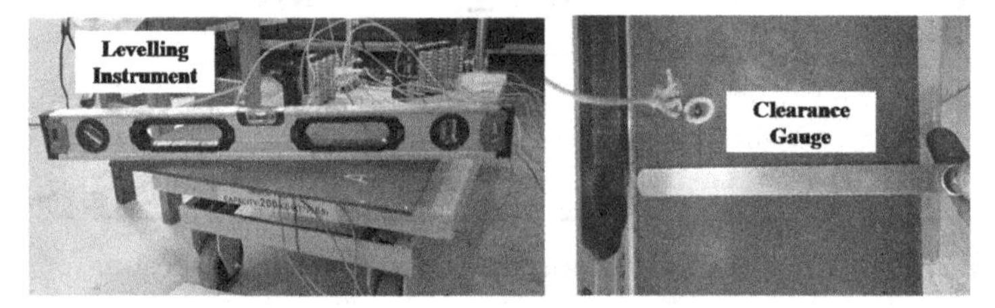

FIGURE 3.5 Planeness testing of steel plate before induction heating

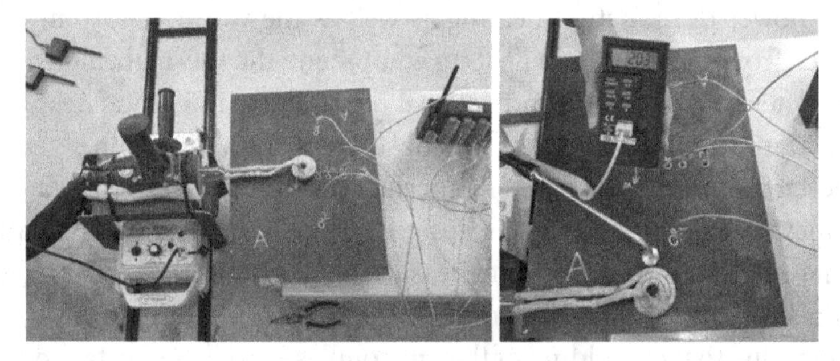

FIGURE 3.6 Experiment process of induction heating

to induction heating when the gap between steel plate and platform is less than 0.5 mm as indicated in Figure 3.5. Then, the next experiment of induction heating with large output power as listed in Table 3.3 can be carried out until the out-of-plane distortion is generated. The actual experiment of induction heating can be seen in Figure 3.6.

3.2.2 Temperature Prediction of Induction Heating

Due to the complexity of coupling analysis of electric magnetic thermal with FE computation, it is not necessary to examine temperature field

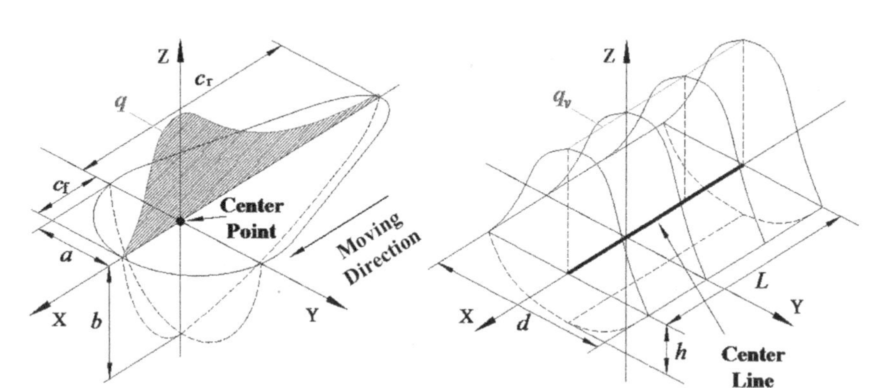

FIGURE 3.7 Body heat source model of thermal analysis

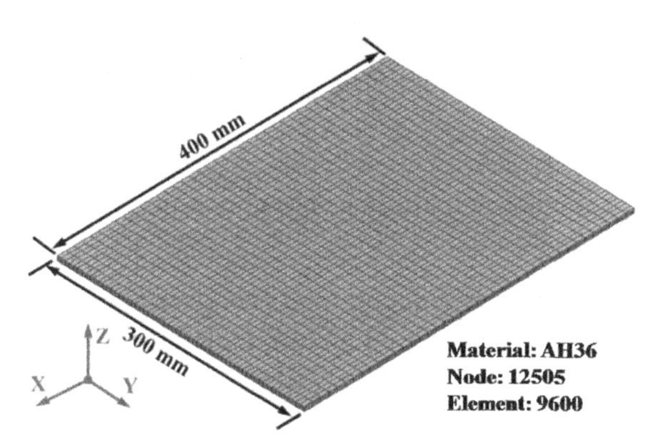

FIGURE 3.8 Solid elements FE model of steel plate with induction heating

of induction heating when the equivalent and simplified heat source model are employed to concentrate on transient feature of temperature profile.

Because of heat generation in the steel plate by induction heating, body heat source model, as shown in Figure 3.7, is employed to represent the thermal response caused by induction heating. Finite element model of examined steel plate as illustrated in Figure 3.8 was created by brick solid element with eight nodes, which has identical geometry and dimension compared with experimental steel plate with induction heating. In addition, there are 12,505 nodes and 9,600 elements of finite element mesh to examine thermal response due to induction heating on AH36 steel plate.

TABLE 3.4 Computational parameters of simple body heat source for induction heating.

Parameters	a	b	c
Value	36 mm	36 mm	5 mm

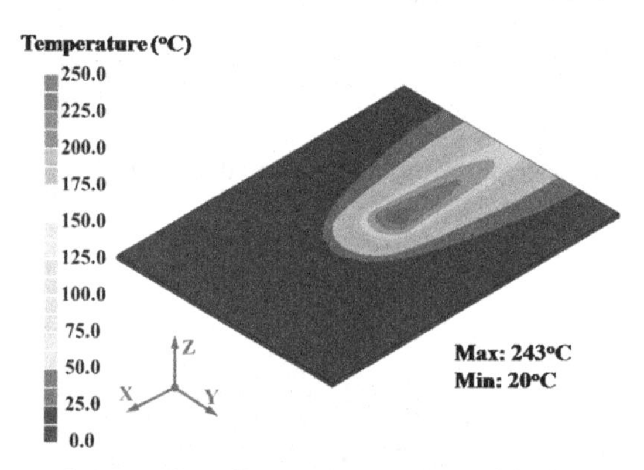

FIGURE 3.9 Distribution of transient temperature with induction heating

During the temperature computation, geometrical parameters of body heat source model are determined by induction heating coil and physical feature (penetration depth) of high-frequency induction heating, while its length and width are both about 36 mm, and its depth is about 5 mm. As listed in Table 3.4, thermal efficiency of induction heating could be confirmed by comparing the computed thermal cycle and measured data.

With thermal analysis of induction heating, transient temperature distribution can be observed and the maximal temperature under the induction heating coil is about 240°C as shown in Figure 3.9, while the moving speed of induction heating coil is 5.0 mm/s and output current is 550 A. With the comparison between computed results and measurement data as demonstrated in Figure 3.10, not only the tendency but also the value of thermal cycle during induction heating has good agreement with each other, while appropriate computed parameters such as material properties and heat source model were invalidated. Furthermore, with the larger heat input of induction heating by increasing output current or decreasing moving speed, higher maximal temperature will be generated.

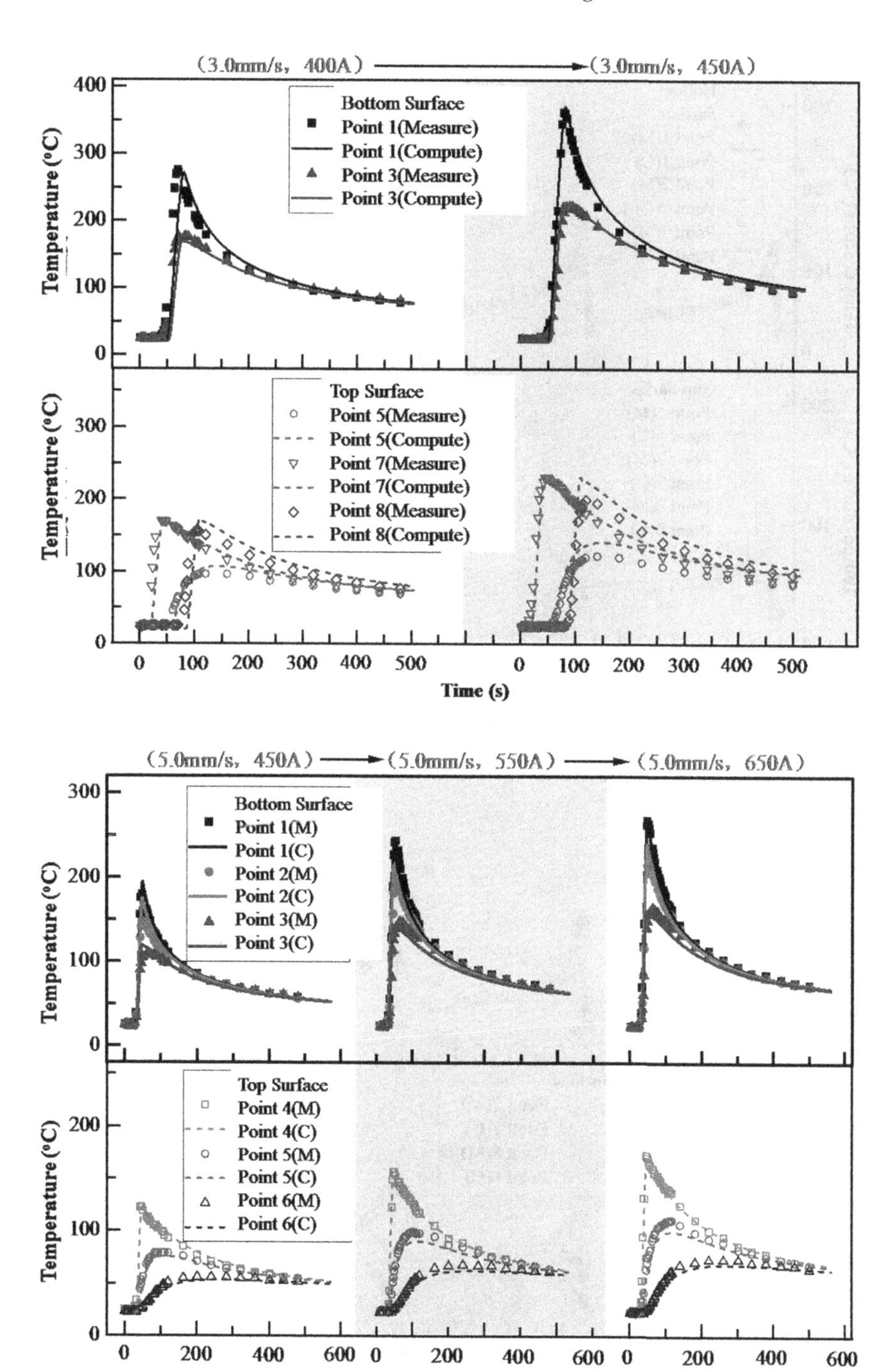

FIGURE 3.10 Comparison of thermal cycles due to induction heating by computed and measured results

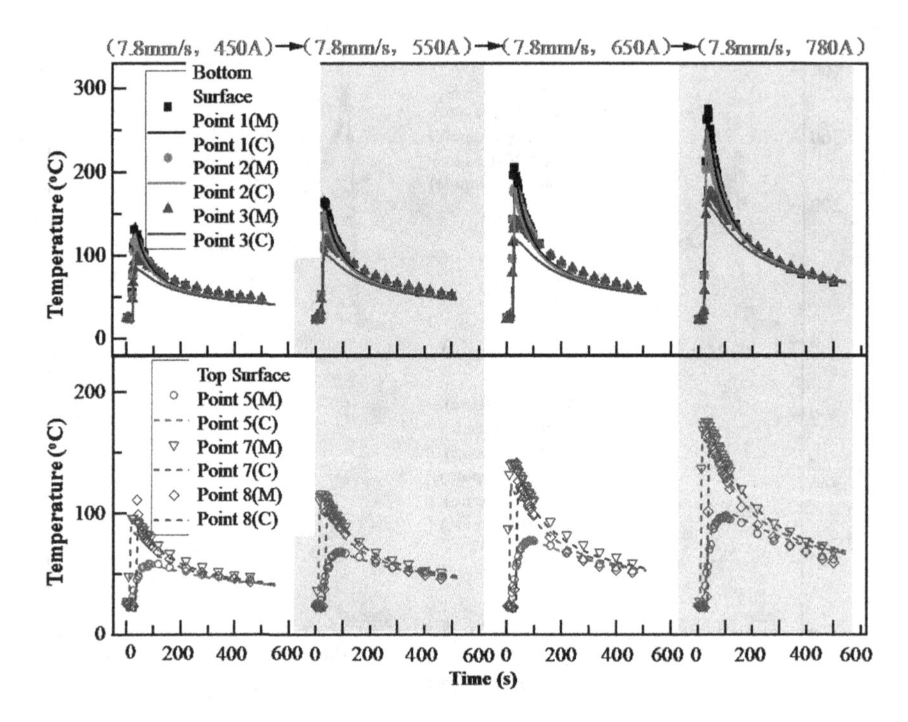

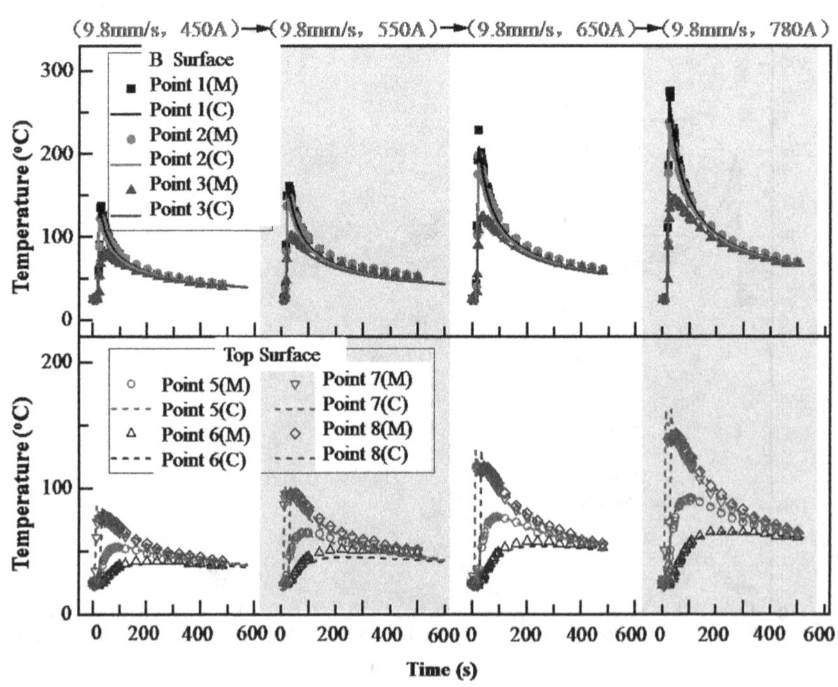

FIGURE 3.10 (Continued)

3.3 GMAW BUTT WELDING EXPERIMENT AND MEASUREMENT

The material of butt welding experiment is ship steel of AH36, which is also a common high tensile strength steel. The dimension of plate for butt-welded joint is 600 mm in length, 200 mm in width, and 5 mm in thickness. In addition, MAG (metal active gas) arc welding method was employed, while the shielding gas is the mixed gas with 80% Ar and 20% CO_2; welding experiment was carried out with NB-350 welding equipment and SB-10-350 wire feeder as demonstrated in Figure 3.11. In actual welding, solid wire of ER50-6 with similar strength compared with ship steel of AH36 was employed, while the diameter of solid wire is 1.2 mm.

In order to penetrate the steel plate during welding, a particular welded groove for butt welding was designed as shown in Figure 3.12, while a welded groove with type of V was employed and the angle of welded groove was 60 degrees. Moreover, there is no gap between two welded plates, and ceramics gasket was employed under the welding bead to achieve double-side forming with single-side welding.

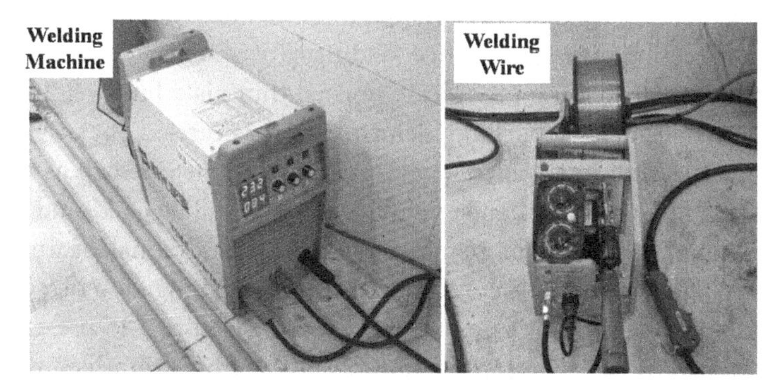

FIGURE 3.11 Equipment for welding experiment

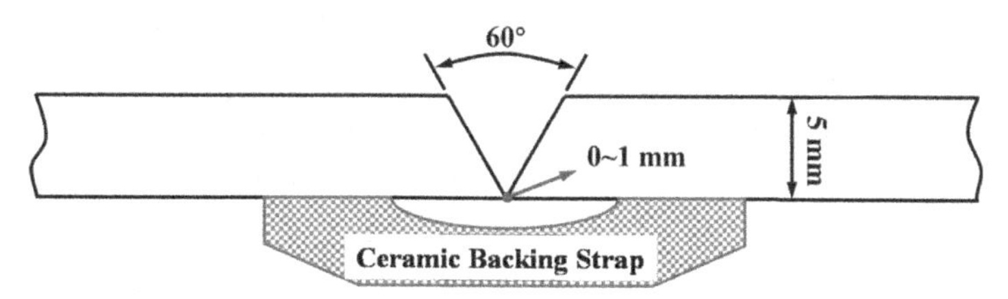

FIGURE 3.12 Design of welded groove for butt welding

3.3.1 Comparative Butt Welding Experiment of Thin Plates

In the preparation stage of butt welding, welded groove was made beforehand. Then, the two welded plates were assembled with tack welding, while there are five tack welding points distributed the backing of welding bead with distance of 150 mm; and grinding machine was employed to polish the excess weld metal of tack welding and ceramics backing strap was fixed as illustrated in Figure 3.13.

Figure 3.14 shows the whole platform of welding experiment and measurement with transient thermal tension, which includes the butt-welded joint, welding equipment, induction heating equipment, temperature measurement device, and data record and storage unit. For the normal butt welding experiment, the established platform can also be employed without the induction heating.

Due to the significant effect of gap between induction heating coil and plate on heat efficiency of induction heating, as examined above, all welding experiment with transient thermal tension was designed with identical gap. In detail, high-temperature resistant tape with material of refrasil glass fiber was employed to fix a ceramics filler piece with thickness of 8 mm under the induction heating coil, which will make sure the constant gap between induction heating coil and steel plate, as well as identical output power of induction heating during the entire welding experiment.

For the actual welding procedure, normal thin plate butt welding with GMAW to consider the buckling deformation was carried out as a benchmark investigation in advance. Then, welding parameters such as welded groove, welding condition, and cooling technique will be fixed, and induction heating with different parameters such as additional heating was considered to achieve transient thermal tension for mitigation of welding-induced buckling of thin plate butt welded joint. Figure 3.15 presents the image of welding buckling mitigation of butt welding with technique of transient thermal tension, in which only one additional heating was employed on the butt welded joint. The diameter of induction heating coil is about 52 mm. The transverse distance d1 between induction heating coil and welding line is about 150 mm, and induction heating coil will move with the identical direction and speed of welding arc, which also means that the longitudinal distance d2 between induction heating coil and welding line is zero.

As shown in Figure 3.16 of butt-welded joint with normal procedure of GMAW, manual welding was used with one experienced welder.

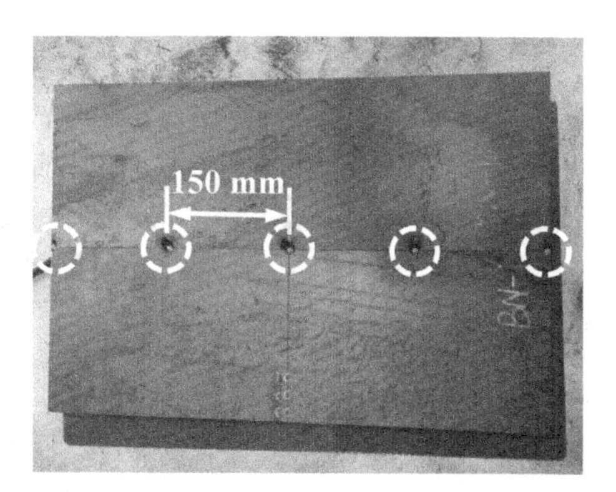

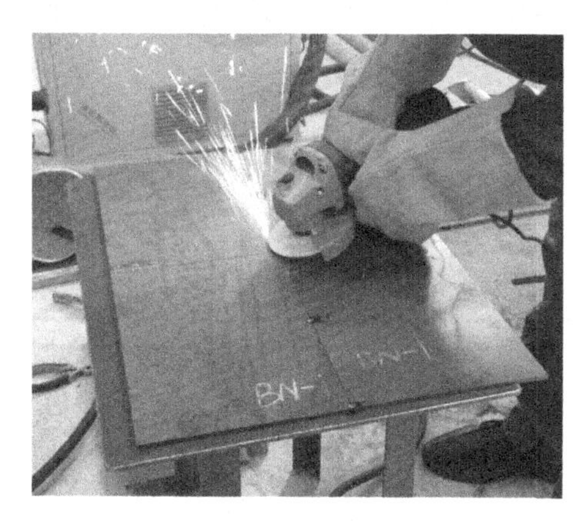

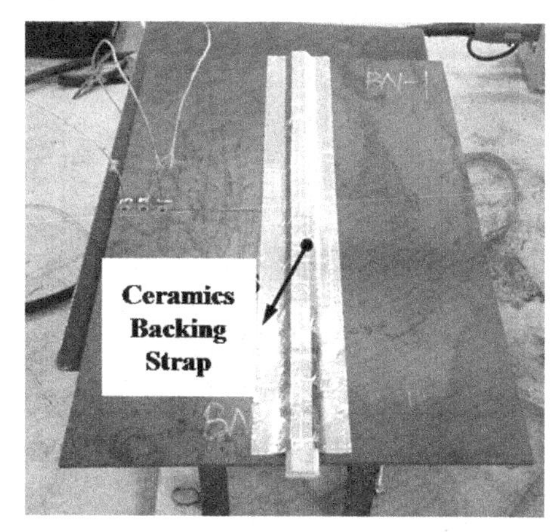

FIGURE 3.13 Preparation of thin plate butt welding

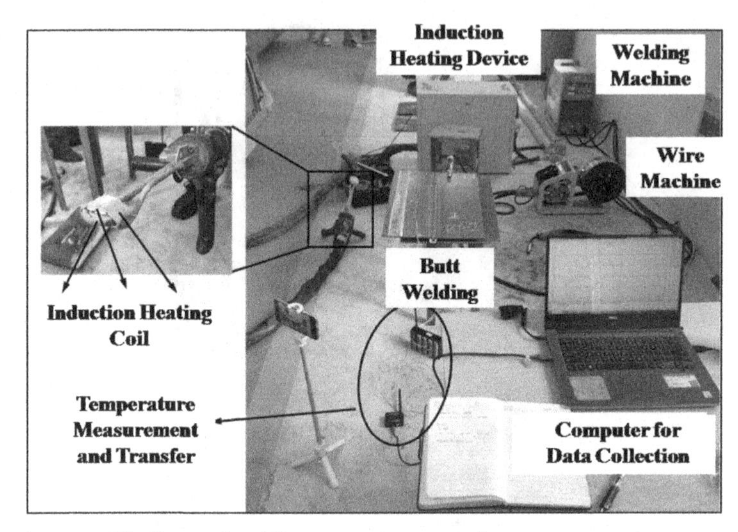

FIGURE 3.14 Platform of welding experiment and measurement.\

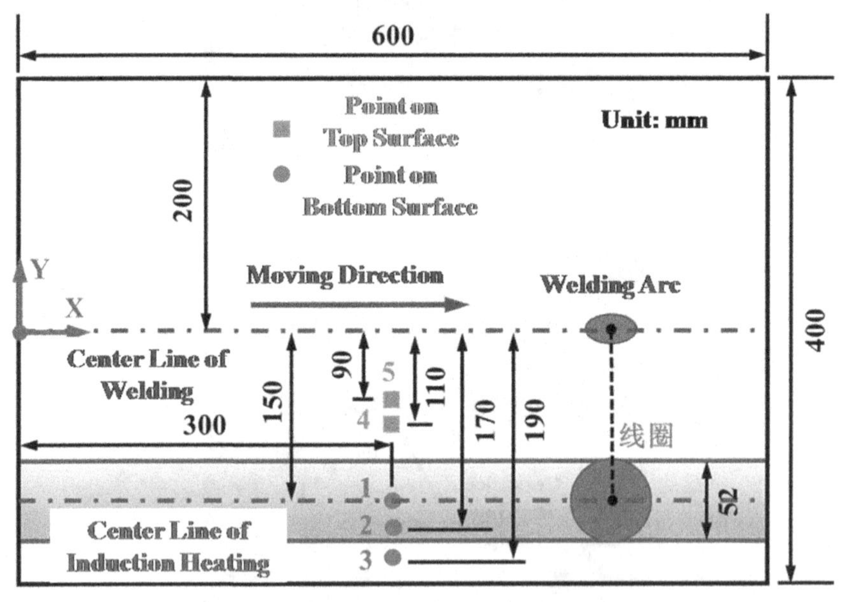

FIGURE 3.15 Image of butt welding with induction heating and location of thermocouples

FIGURE 3.16 Thin plate butt welding with GMAW

During welding procedure with transient thermal tension, there are two experienced welders to carry out the buckling mitigation of thin plate butt welding while one welder will finish the welding according to procedure of normal welding and the other will operate induction heating coil to move with welding arc simultaneously for transient thermal tension.

In order to examine the mitigation of welding buckling distortion of thin plate butt welding, there are three cases of transient thermal tension experiment with different magnitudes of additional induction heating as well as maximal heated temperature. Welding and induction heating conditions are summarized in Table 3.5, while the butt-welded joint with normal welding is entitled BN-1 and three butt-welded joints with transient thermal tension are entitled BT-1, BT-2, and BT-3, respectively. In

TABLE 3.5 Welding conditions and parameters of induction heating during butt welding experiment.

Butt Welding No.	BN-1	BT-1	BT-2	BT-3
Welding Procedure	Normal	With Thermal Tension Due to Induction Heating		
Current (A)	265~270	265~271	256~262	255~260
Voltage (V)	22.2~22.4	23.2~23.4	23.2~23.4	23.2~23.3
Welding Time (s)	140	143	142	138
Moving Speed (mm/s)	4.3	4.2	4.2	4.3
Current of Induction Heating (A)	/	450	475	500
Transverse Distance d1 (mm)	/	150	150	150
Longitudinal Distance d2 (mm)	/	0	0	0

addition, with the increment of output power of induction heating, the magnitude of additional heating source as well as maximal temperature will be increased.

Due to the manual operation of welding experiments, there is a tiny difference in welding condition such as moving speed. In general, the welding conditions such as current, voltage, and moving speed can be considered identical for welding procedure with and without thermal tension due to induction heating. As summarized in Table 3.5, there are welding conditions and parameters of induction heating, which will be employed for the later thermal elastic plastic FE computation.

During butt welding with thermal tension caused by induction heating, K-type thermocouple of platinum-rhodium alloy and temperature measurement system of JM3818A as mentioned above were employed again to obtain the transient temperature of several points on examined butt-welded joint, while the temperature profile due to welding arc and induction heating will also be investigated. In addition, measured points of transient temperature were designed as indicated in Figure 3.17 to avoid the moving effect of induction heating coil. There are two measured points on the top surface and three measured points on the back surface of examined butt welded joints. Moreover, measured points 1 and 2 are located at back surface just under the induction heating coil to measure the transient temperature of heating zone due to induction heating, and measured points 4 and 5 are located at the region of top surface between welding line and area of induction heating.

K-type thermocouple of platinum-rhodium alloy was fixed on the top and back surfaces with professional equipment as shown in Figure 3.17.

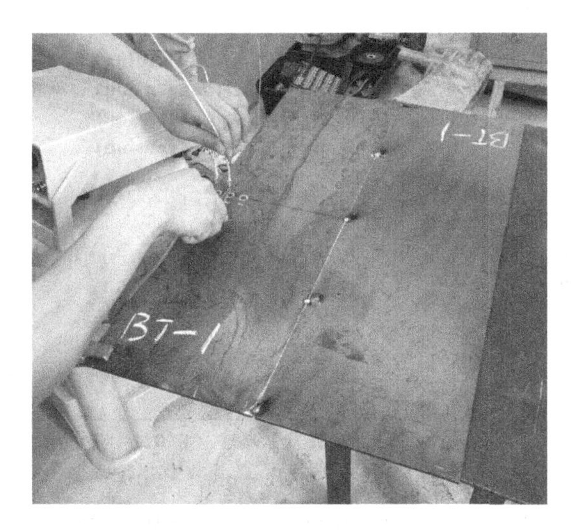

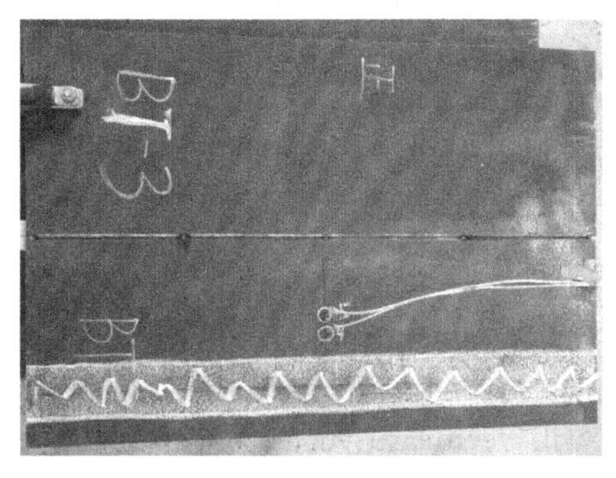

FIGURE 3.17 Positions of thermocouples and their installation

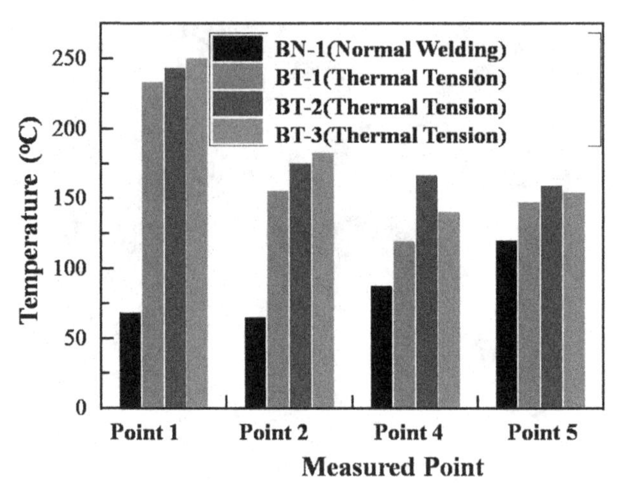

FIGURE 3.18 Comparison of maximal temperature of measured points on butt-welded joints

Measured maximal temperatures of considered points during GMAW welding of butt-welded joints without and with thermal tension were compared as indicated in Figure 3.18. It can be seen that maximal temperatures of measured points 1 and 2 located at the region with induction heating are obviously higher than that without induction heating, and maximal temperature of measured points 1 and 2 will increase according to the increasing output current as well as output power of induction heating. Besides, maximal temperatures of measured points 4 and 5 located at the region between welding line and area of induction heating have almost identical values compared with and without induction heating, while induction heating has neglectable effect on temperature field caused by welding.

After welding experiment and cooling down to room temperature, there are four butt welded joints as demonstrated in Figure 3.19, while they more or less have out-of-plane welding distortion, in particular welding buckling.

3.3.2 Measurement of Butt-Welded Joints after Cooling Down

After cooling down to room temperature, all butt-welded joints as mentioned above will be measured for out-of-plane welding distortion and test of metallographic phase sequentially. For precious measurement

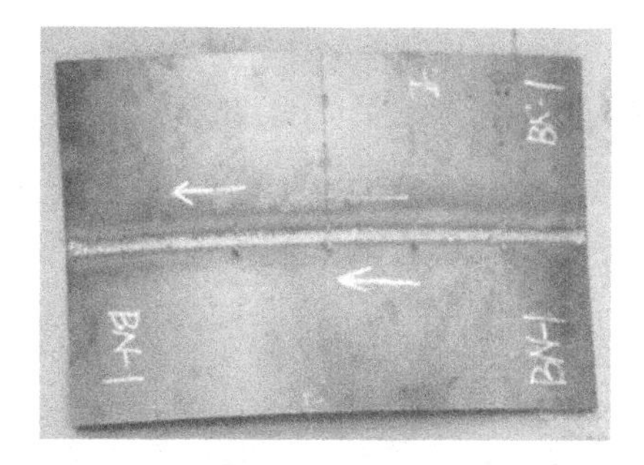

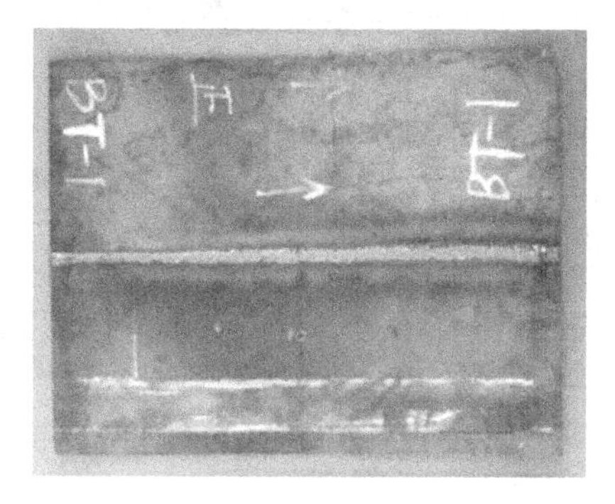

FIGURE 3.19 Butt-welded joints with different procedures

FIGURE 3.20 Three-coordinate measured machine supported by Hexagon Company

of out-of-plane welding distortion, three coordinate measured machine with Global Plus 686 (measured accuracy: 0.1 μm) supported by Hexagon company as shown in Figure 3.20, was employed, while its measured range is 800 mm in longitudinal direction, 600 mm in transverse direction and 600mm in thickness direction.

In addition, measured probe is controlled by computer with the regular route designed beforehand; each measured point will be individually considered and corresponding measured software will show the measured result as welding distortion immediately based on the coordinates of measured point cloud. As indicated in Figure 3.21, distances between adjacent measured points are 30 mm and 20 mm in the longitudinal and the transverse directions, respectively. This kind of measured plan of out-of-plane welding distortion for butt-welded joints is determined by measured accuracy and cost.

In general, steel rust as well as welding spatter on the measured surface of experimental butt-welded joints should be removed and polished to guarantee the measured accuracy before welding distortion measurement. Then, the butt-welded joint was sequentially placed on the operation

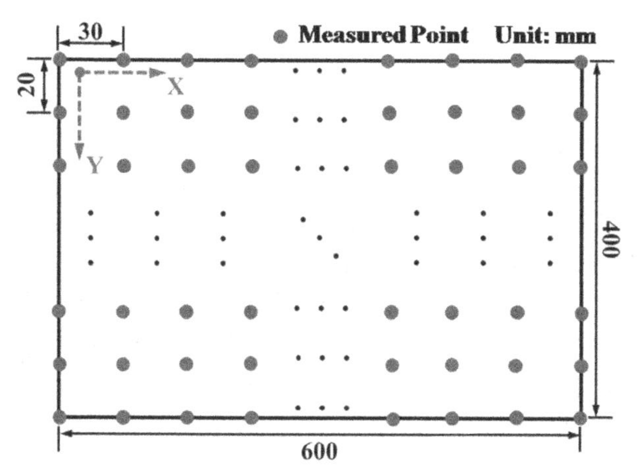

FIGURE 3.21 Measured plan of out-of-plane welding distortion of butt-welded joints

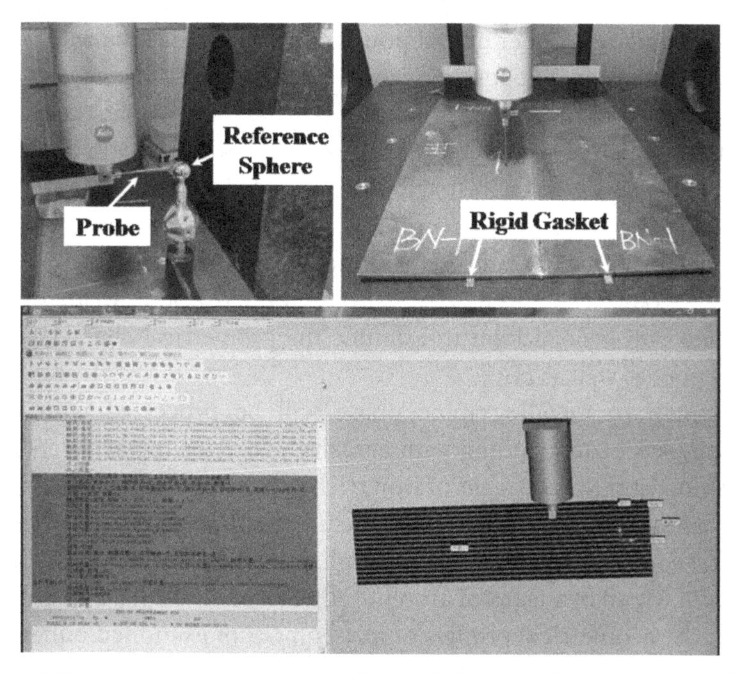

FIGURE 3.22 Measurement process of out-of-plane welding distortion of butt-welded joint

platform of three-coordinate measured machine, which should be fixed to prevent movement during the measurement process. As demonstrated in Figure 3.22, the entire measurement processes of out-of-plane welding distortion of butt-welded joints were presented.

Since in-plane shrinkage distortion can be ignored compared with out-of-plane welding distortion for examined thin plate butt-welded joints, out-of-plane welding distortion will be paid more attention in the later investigation. After measurement, the coordinates of point cloud will be dealt with post process for contour plotting of out-of-plane welding distortion as shown in Figure 3.23. It can be seen that four examined butt-welded joints with AH36 thin plate are all deformed with saddle mode, which is the typical feature of welding-induced buckling. In addition, butt-welded joint will be bent in different directions in longitudinal and transverse directions. For the butt welded joint of BN-1, it bends upward in the welding line direction and bends downward in the direction perpendicular to welding line.

Compared with the butt-welded joint of BN-1, butt-welded joints such as BT-1, BT-2, and BT-3 with thermal tension due to induction heating as examined above have less magnitude of out-of-plane welding distortion, and moreover, larger output power of induction heating with welding experiments has much better reduction effect of out-of-plane welding distortion. Taking the butt welded joint of BT-3 as an example, the maximal deflection, which is considered as difference between maximal and minimal out-of-plane welding distortion, was reduced from 23.88 mm without thermal tension to 12.17 mm.

By means of wire-electrode cutting with spark erosion, specimen of metallography measurement was made in the region of welded zone of examined butt-welded joint to examine the geometrical characteristic of molten pool as well as HAZ.

Specimen of metallographic phase will be processed with coarse grinding, precision grinding with abrasive paper for metallography, and burnishing, and then reagent of iron trichloride was employed to corrode the cross-section of specimen of metallographic phase. After several minutes, specimen of metallographic phase will be washed with distilled water and wiped by means of absolute ethyl alcohol. With Drying with air blower, the geometrical profiles of molten pool of examined butt-welded joints can be eventually observed as shown in Figure 3.24. It can be seen that all butt-welded joints with AH36 thin plate were jointed with full penetration, and a tiny difference at bottom region of molten pool shape also can be observed which may come from the influence of manual welding and installation tolerance of ceramics backing strap. However,

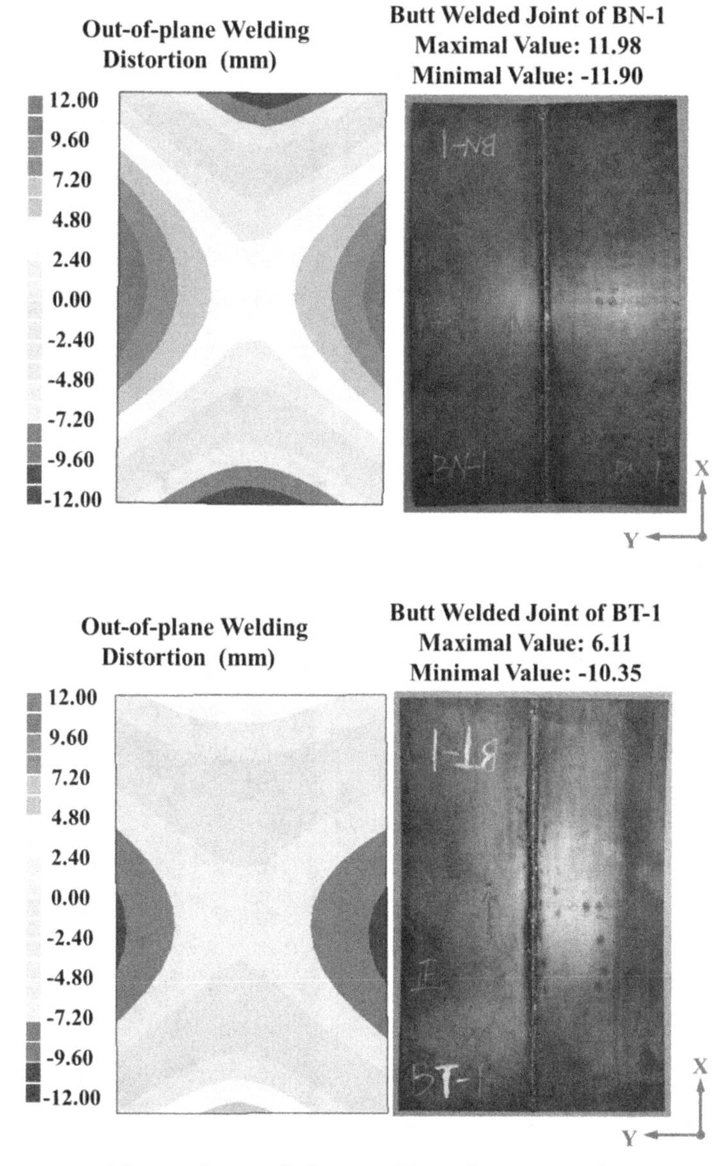

FIGURE 3.23 Measured out-of-plane welding distortion of examined butt-welded joints

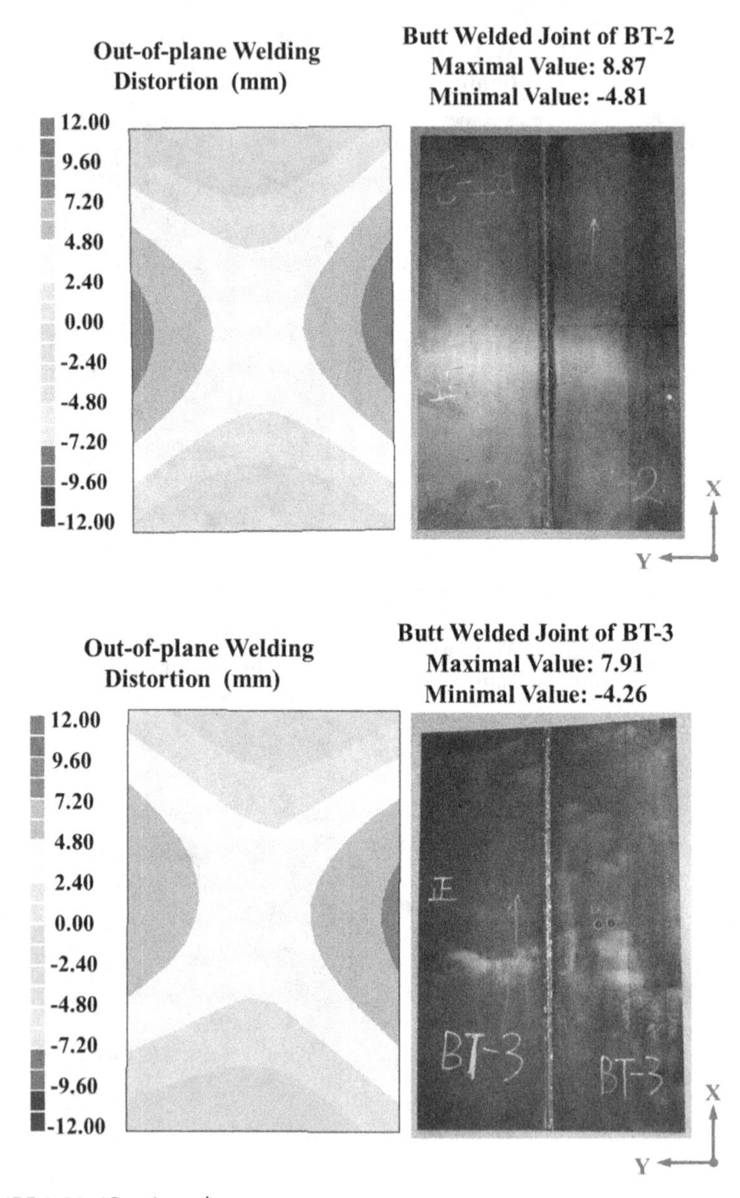

FIGURE 3.23 (Continued)

overall molten pools of all examined butt-welded joints are almost identical, which can confirm that welding conditions of butt-welded joint experiments without and with thermal tension caused by induction heating should be consistent.

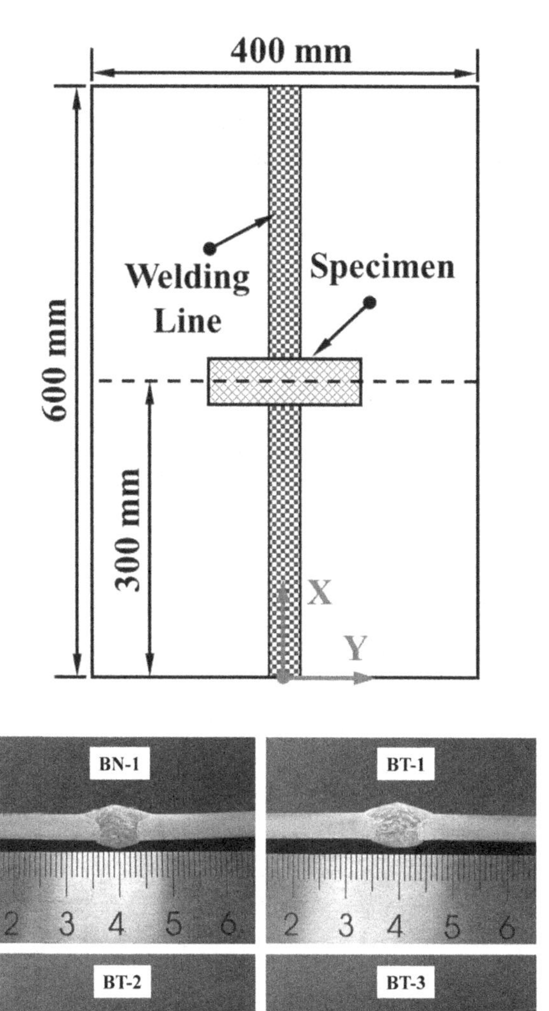

FIGURE 3.24 Result of metallographic phase of butt-welded joints

3.4 SUMMARY

Thin plates with AH36 were employed for butt-welded fabrication, while normal welding procedure and welding with transient thermal tension by induction heating were both considered. Computational parameters of induction heating were confirmed by means of experiment and FE computation beforehand; thermal cycles during induction heating and welding were both measured and compared. Not only the out-of-plane welding distortions with three coordinate measured machine but also metallographic phases were measured and compared with different welding procedures.

CHAPTER 4

Butt Welding Buckling Prediction with FE Analysis

WITH THE PREVIOUS WELDING experiment and measurement, buckling behavior due to welding with its particular features was observed; meanwhile, transient thermal tension with induction heating was also employed to reduce the welding-induced buckling, while induction heating coil will move simultaneously with welding arc. Significant differences in butt-welded joints with conventional welding and welding with transient thermal tension can be observed from the measurement results. Moreover, the transient thermal elastic plastic FE computation will be carried out to represent the thermal and mechanical responses during the welding process in this section, and elementary cause as well as mechanical mechanism will also be clarified based on the computed results.

4.1 SOLID ELEMENT MODEL AND MATERIAL PROPERTIES

Taking the balance of computational cost and comparison of thermal elastic plastic FE computation, benchmark based on molten pool shape and welding conditions of butt-welded joint with conventional welding (BN-1) was considered to investigate all butt welding experiments without and with transient thermal tension.

Solid elements FE model of examined butt-welded joint was then made as shown in Figure 4.1, while the geometrical profile of molten pool was

DOI: 10.1201/9781003442523-4

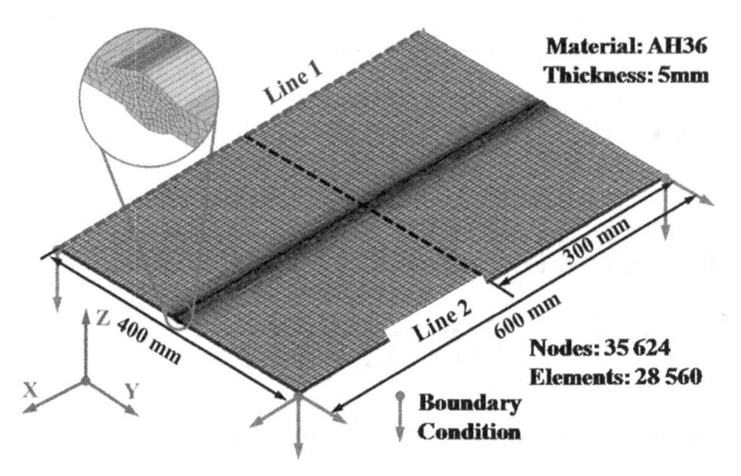

FIGURE 4.1 Solid elements model of examined butt welded joint

determined by the metallographic phase of butt-welded joint with conventional welding (BN-1) as demonstrated in Figure 4.1. In addition, the employed solid element is a typical brick element with eight nodes, and the element length in the welding line is 10 mm. As shown in Figure 4.1, the region with high temperature near the welding line is usually meshed with much fine mesh to accurately capture thermal and mechanical response by means of transient thermal elastic plastic FE computation. Meanwhile, regions far away from the welding line such as base material were also meshed with a little bit of fine mesh, which will be employed to consider additional heating caused by induction heating as transient thermal tension. This kind of element mesh for examined butt-welded joint could guarantee computational accuracy and reduce the computing time.

In addition, number of nodes and elements in the solid elements FE model of examined butt-welded joint as shown in Figure 4.1 are 35,624 and 28,560, respectively. Points on Line 1 (along the welding line) and Line 2 (normal to the welding line) were also selected to compare the computed and measured out-of-plane welding distortion later.

4.2 HEAT TRANSFER AND TEMPERATURE VALIDATION

During the thermal analysis, welding arc and additional heating with induction heating will be simulated by body heat source model, while uniform density distribution of heat flux will be employed as heating input. The density magnitude of heat flux can be obtained by welding conditions such as current, voltage, and speed. In addition, shape parameters of body

TABLE 4.1 Parameters of heat source model of butt-welded joint.

Shape Parameter of Heat Source Model	a	b	c
Value (Welding)	15 mm	7.8 mm	5.2 mm
Value (Induction Heating)	36 mm	36 mm	5 mm

heat source for additional heating by induction heating was examined in advance, and shape parameters of body heat source for welding can be confirmed based on the metallographic phase of butt-welded joints as summarized in Table 4.1. The initial temperature and room temperature are both considered to be 20°C and the nodes on the surface of FE model will lose the heat by means of heat convection and heat radiation, which is also considered as thermal boundary condition. The time increment is 0.2 seconds during the welding, and after welding with cooling, the time increment will be increased by means of exponential form to enhance the computational efficiency.

Due to the design of equal strength match of examined butt-welded joints, the base material and filler metal have almost identical thermal and mechanical properties; thus, temperature-dependent material properties as shown in Figure 4.2 will be employed during the FE computation.

As shown in Figure 4.3, computed temperature distribution was demonstrated and compared when the welding arc is passing the center of butt-welded joint. It can be seen that temperature variation only exists near the welding line for conventional welding, and there is both temperature variation in the region of welding line and induction heating with the technique of transient thermal tension. In addition, the temperature gradient in the region at the front of welding heat source is much larger than that in the rare region of welding heat source; temperature caused by welding is also much larger than that caused by induction heating.

Distribution of maximal temperature of points on the middle cross-section can be obtained and compared for different welding procedures as shown in Figure 4.4. Compared with conventional welding, the welding process with thermal tension will only change the temperature for the region with induction heating, and there is not any influence on original welding temperature. The maximal temperature due to induction heating can be confirmed to be 237°C, 247°C, and 258°C for butt-welded joints of BT-1, BT-2, and BT-3, respectively.

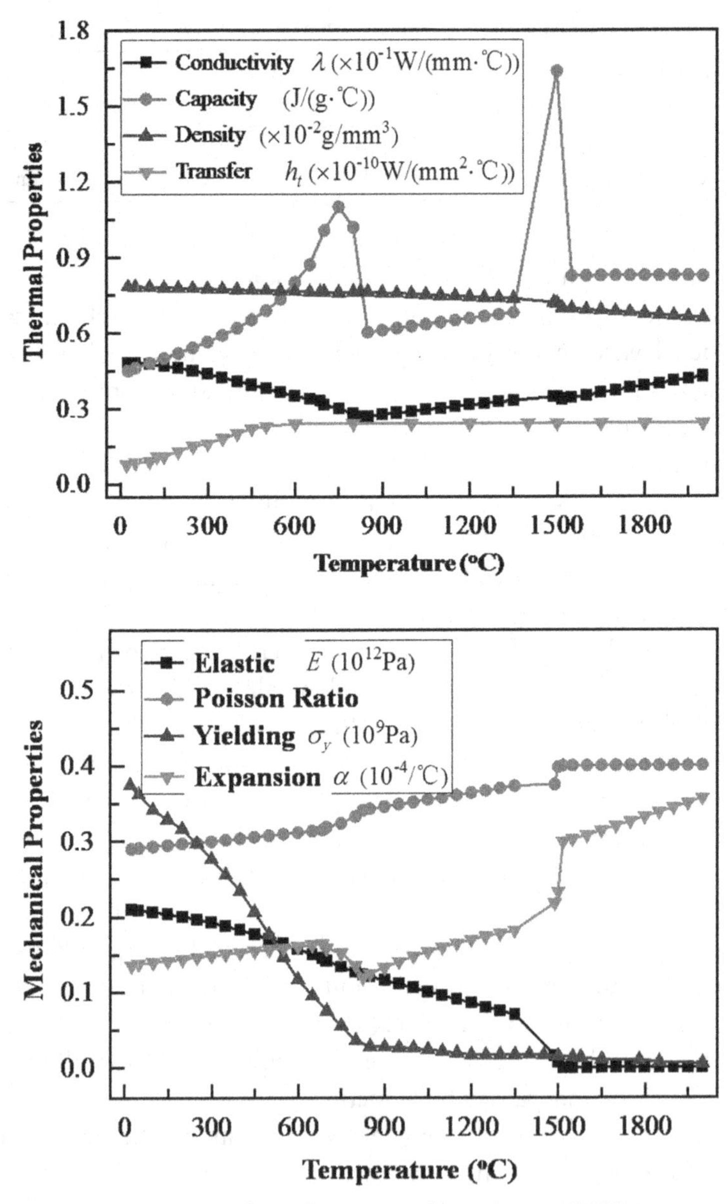

FIGURE 4.2 Temperature-dependent material properties of AH36

In order to confirm the computed accuracy of thermal analysis, shape of molten pool as well as thermal cycles were then considered and compared as demonstrated in Figures 4.5 and 4.6. In Figure 4.5, there is a good agreement of molten pool between the computed results and measurement.

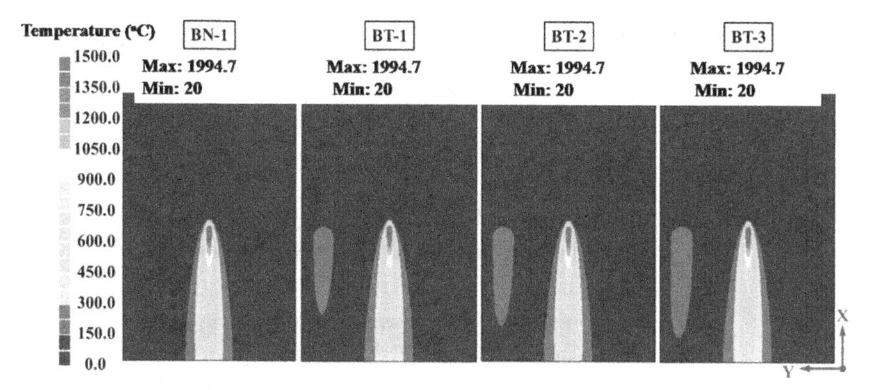

FIGURE 4.3 Transient temperature contour of butt-welded joint without and with thermal tension

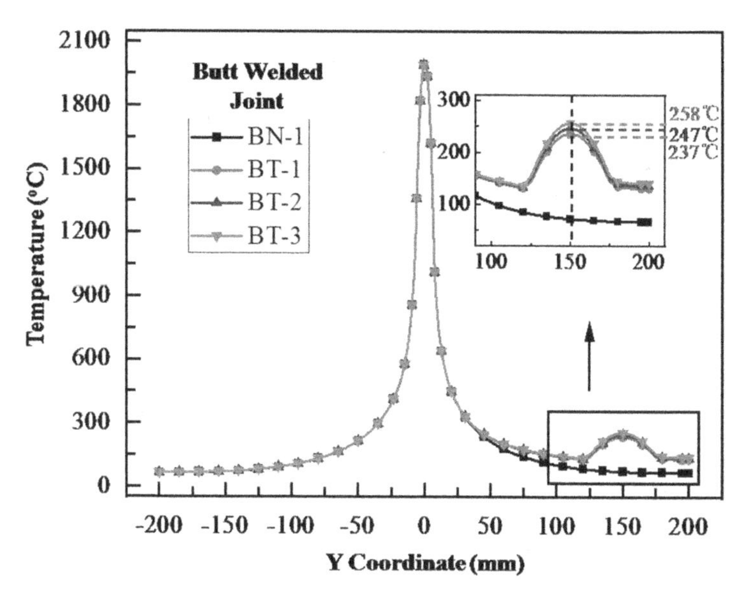

FIGURE 4.4 Distribution comparison of maximal temperature of butt-welded joint

Moreover, the measured points of No. 1 and No. 2 were located in the region near welding line, and the measured points of No. 4 and No. 5 were located in the region between welding line and induction heating. Good agreement of thermal cycles of considered points during conventional welding and welding with induction heating also can be observed as demonstrated in Figure 4.6. The little difference could be caused by measured tolerance due

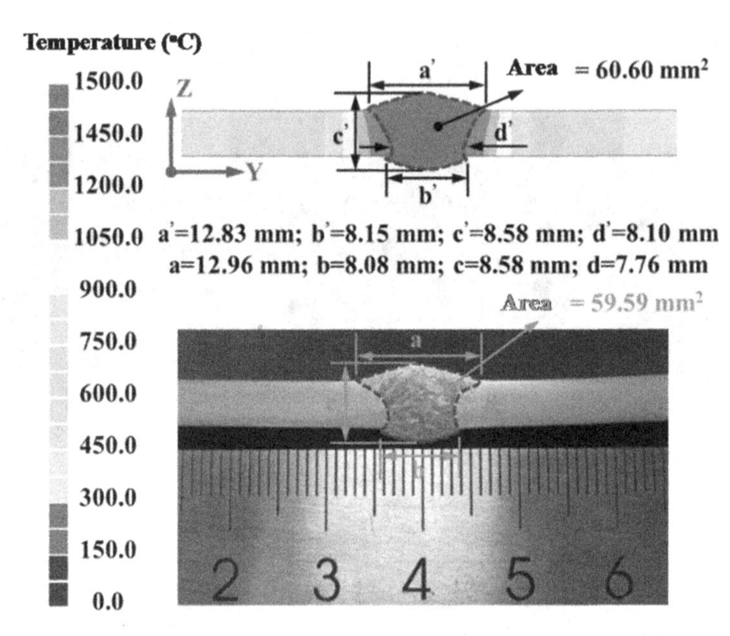

FIGURE 4.5 Comparison of molten pool of butt-welded joint

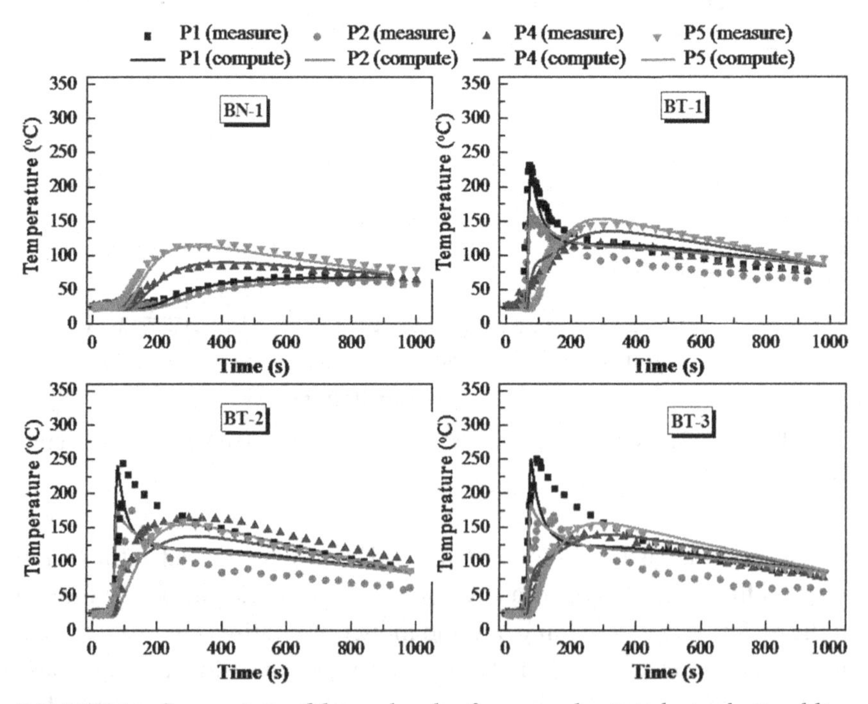

FIGURE 4.6 Comparison of thermal cycle of measured points during butt welding

to manual welding, thermocouple tack welding, position of thermocouple, and computational tolerance such as model simplification and material properties at high temperature.

Overall, reliable thermal analysis with FE computation can be concluded with the maximal temperature and its variation tendency, and coupled temperature was accurately examined for the welding and induction heating simultaneously.

4.3 WELDING BUCKLING PREDICTION WITH FINITE STRAIN

Taking the previously computed temperature of nodes as thermal loading to the FE model with solid elements as shown in Figure 4.1, mechanical analysis was then carried out. In addition, Hooke law will be considered during the elastic stage, and Mises yield criterion will be employed for plastic strain computation. The temperature-dependent material properties of AH36, so-called material nonlinearity as indicated in Figure 4.2, were also considered. Moreover, the creep behavior of high temperature as well as solid phase transformation during welding will not be considered due to their less influence and magnitude of generation strain.

For the welding buckling of stability behavior, there are two options for out-of-plane welding distortion prediction of butt-welded joint with AH36 thin plate, which are small deformation theory (infinitesimal strain) and large deformation theory (finite strain) for investigation of geometrical nonlinearity of welding buckling.

For the previous butt welding experiment with normal welding and with transient thermal tension, there are both no external constraints to be free condition, and the rigid body motion of examined butt-welded joint with AH36 will be prevented to be mechanical boundary condition as indicated in Figure 4.1. Compared with in-plane shrinkage distortion, out-of-plane welding distortion as well as welding buckling will be paid more attention later due to its large magnitude when welding buckling occurs. As shown in Figure 4.7, computed out-of-plane welding distortion of butt-welded joint of AH36 thin plate can be observed, while the deformed rate is 5. It can be seen that contour of computed out-of-plane welding distortion with large deformation theory has a similar deformed tendency (saddle deformation mode) and magnitude compared with measured results, and there are obvious differences between computed contour of out-of-plane welding distortion with small deformation theory and measurement. The difference

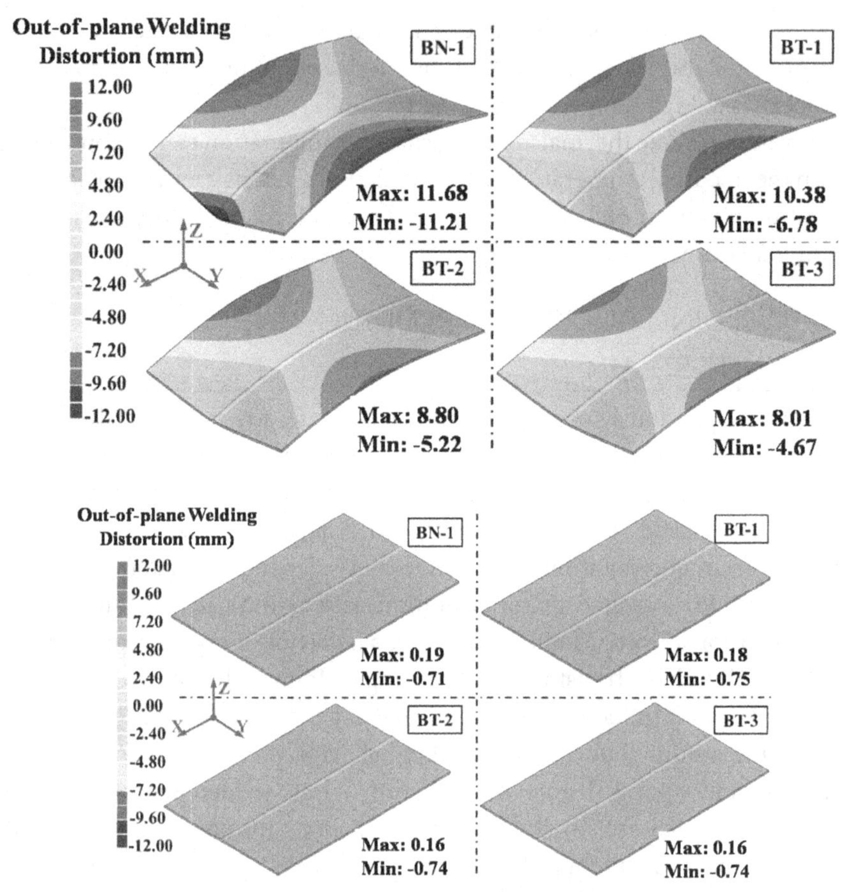

FIGURE 4.7 Computed contour of out-of-plane welding distortion of butt-welded joint of AH36 thin plate (Deformed rate: 5)

between computed out-of-plane welding distortions with small and large deformation theories can confirm that examined butt-welded joint of AH36 thin plate buckled due to welding heat input.

In order to closely examine the computed results, out-of-plane welding distortions of points on line 1 and line 2 as indicated in Figure 4.1 were considered and compared as demonstrated in Figure 4.8. Moreover, with the large deformation theory, computed results of out-of-plane welding distortion have good agreement compared to measurement data with welding buckling features. The points on line 1 along welding line have upward bending deformation, and the points on line 2 normal to the welding line bend downward, which have opposite deformed tendency to be saddle type of typical welding buckling.

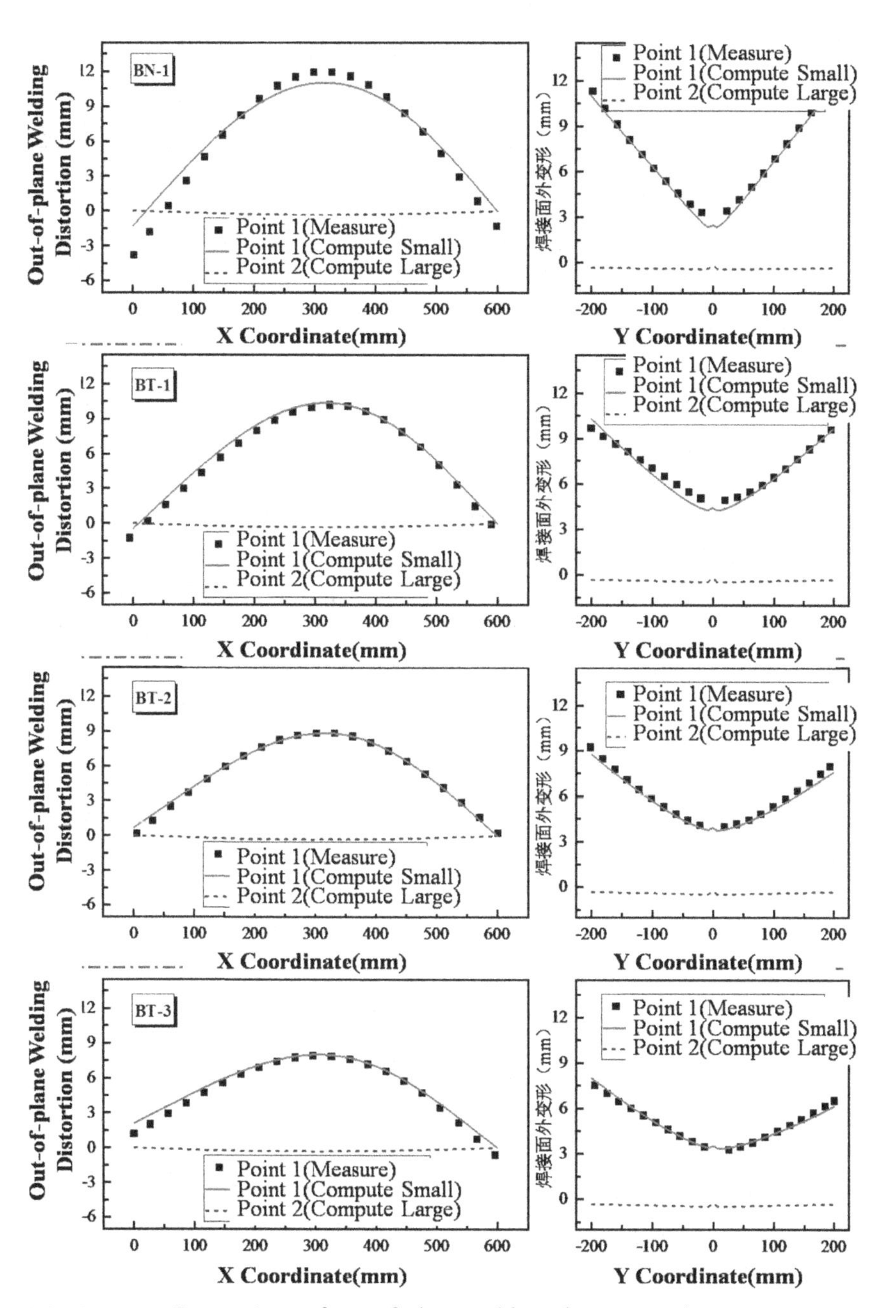

FIGURE 4.8 Comparison of out-of-plane welding distortion of points on Line 1 and Line 2 of butt-welded joint

TABLE 4.2 Tolerance comparison of computed and measured results of examined butt-welded joint.

		Large Deformation Theory		Small Deformation Theory	
No	**Measurement**	**Computation**	**Tolerance**	**Computation**	**Tolerance**
BN-1	23.88 mm	22.89 mm	4.1%	0.90 mm	96.2%
BT-1	16.46 mm	17.16 mm	4.3%	0.93 mm	94.3%
BT-2	13.68 mm	14.02 mm	2.5%	0.90 mm	93.4%
BT-3	12.17 mm	12.68 mm	4.2%	0.90 mm	92.6%

When the small deformation theory is considered, computed out-of-plane welding distortion is much less, which has a large difference compared with measured results. Considering the maximal relative out-of-plane welding distortion, which is defined as difference value of maximal and minimal out-of-plane welding distortion, the tolerances between computed results and measured data were analyzed and summarized in Table 4.2. There is only 5% tolerance between computed results and measured data when the large deformation theory as well as finite strain theory is employed, and FE computation can accurately predict out-of-plane welding distortion of examined butt-welded joint with conventional welding and welding associated by induction heating as thermal tension. Meanwhile, computed tolerance is much larger by comparing predicted results and measurements when small deformation theory is employed.

Moreover, the previous computation also demonstrated that welding buckling behavior of butt-welded joint of AH36 thin plate could be represented by FE computation by considering both material nonlinearity and geometrical nonlinearity together, and the employed thermal elastic plastic FE computation solver will not only represent the welding buckling behavior but also predict the magnitude of out-of-plane welding distortion with high accuracy.

Besides, thermal tension in the welding experiment could reduce the out-of-plane welding distortion of butt-welded joint, and out-of-plane welding distortion will be obviously reduced with the increasing output power of induction heating as well as induction temperature. For example, butt-welded joint of BT-3 with the transient thermal tension has the lowest out-of-plane welding distortion, while the maximal out-of-plane welding distortion was reduced by about 50% compared with that with conventional welding.

TABLE 4.3 Comparison of computational time of mechanical analysis (Nodes: 35,624; Elements: 28,560).

	Small Deformation Theory	Large Deformation Theory
Normal Welding	4 hours 5 minutes 8 seconds	35 hours 40 minutes 15 seconds
Thermal Tension with Temperature of 243°C	4 hours 14 minutes 13 seconds	30 hours 49 minutes 47 seconds
Thermal Tension with Temperature of 253°C	4 hours 14 minutes 36 seconds	33 hours 16 minutes 50 seconds
Thermal Tension with Temperature of 278°C	4 hours 7 minutes 9 seconds	30 hours 17 minutes 52 seconds
Thermal Tension with Temperature of 357°C	4 hours 27 minutes 59 seconds	34 hours 23 minutes 46 seconds

For the above thermal elastic plastic FE computation, Dell T7910 Server as well as Ubuntu 16.04 OS platform was employed, while there are 2 CPUs (Intel(R) Xeon(R) Gold 6136 CPU @ 3.00 GHz) with 48 cores and the computational memory is about 196 GB. For the thermal and mechanical analysis, only four physical cores for one computational case will be used with the parallel computation technology. Due to large time consumption of mechanical analysis, computational time to predict out-of-plane welding distortion as well as buckling deformation for the welded joints with normal welding and transient thermal tension (maximal temperature of heated region with induction heating: 247°C, 258°C, 278°C, 357°C) were summarized and compared in Table 4.3.

It can be found that mechanical analysis with large deformation theory to consider the geometrical nonlinearity will consume much more commutating time compared with that with small deformation theory; however, computed results with small deformation theory have obvious differences compared with measured results.

4.4 MECHANISM BASED ON INHERENT DEFORMATION

With the above investigation, transient thermal tension technique with no generation of additional plastic strain in the heated region of induction heating will effectively mitigate the welding-induced buckling of butt-welded joint of AH36 thin plate. There are large differences between computed out-of-plane welding distortion by thermal elastic plastic FE computation with small deformation and large deformation theory.

4.4.1 Comparison of Welding Plastic Strain

Generally, out-of-plane welding distortion is basically determined by residual plastic strain, and the distribution and magnitude of plastic strain will significantly influence the computed accuracy of welding buckling. Thus, the influence of induction heating on plastic strain in the region near the welding line will be examined later.

While there are four considered cases with different maximal temperatures in the heated region (247°C for BT-2, 258°C for BT-3, 278°C, 357°C). Contour plotting of longitudinal and transverse plastic strain of examined butt-welded joint could be obtained by FE computation with small and large deformation theory. As shown in Figure 4.9, additional plastic strain was generated when the additional maximal temperature is about 357°C, and there will be nothing in the heated region with other additional maximal temperatures.

Moreover, there will be a uniform distribution of both longitudinal and transverse plastic strains along the welding line direction without considering the end effect of welding such as welding beginning and completion. The region with longitudinal compressive plastic strain has larger width compared with that of transverse compressive plastic strain. Distributed curve of both longitudinal and transverse plastic strains on middle cross-section was then examined as indicated in Figure 4.10. It can be seen that magnitude of transverse plastic strain is much larger than that of longitudinal plastic strain, while region of transverse plastic strain is much narrower than that of longitudinal plastic strain. This kind of unique distribution is generated by different self-constraints supported by surrounding base material in the longitudinal and transverse directions.

As shown in Figure 4.11, compared with normal welding, longitudinal and transverse plastic strains will both be decreased with the transient thermal tension technique. In addition, computed results of thermal elastic plastic FE computation with small and large deformation theory have almost identical longitudinal and transverse plastic strain for normal welding and transient thermal tension technique.

Therefore, identical mechanical loading as well as plastic strain can be observed during the thermal elastic plastic FE computation with small and large deformation, and the differences of predicted out-of-plane welding distortion as well as welding buckling of FE computation with small and large deformation can be understood with relation of plastic strain and displacement, which is considered geometrical governing equation. Welding

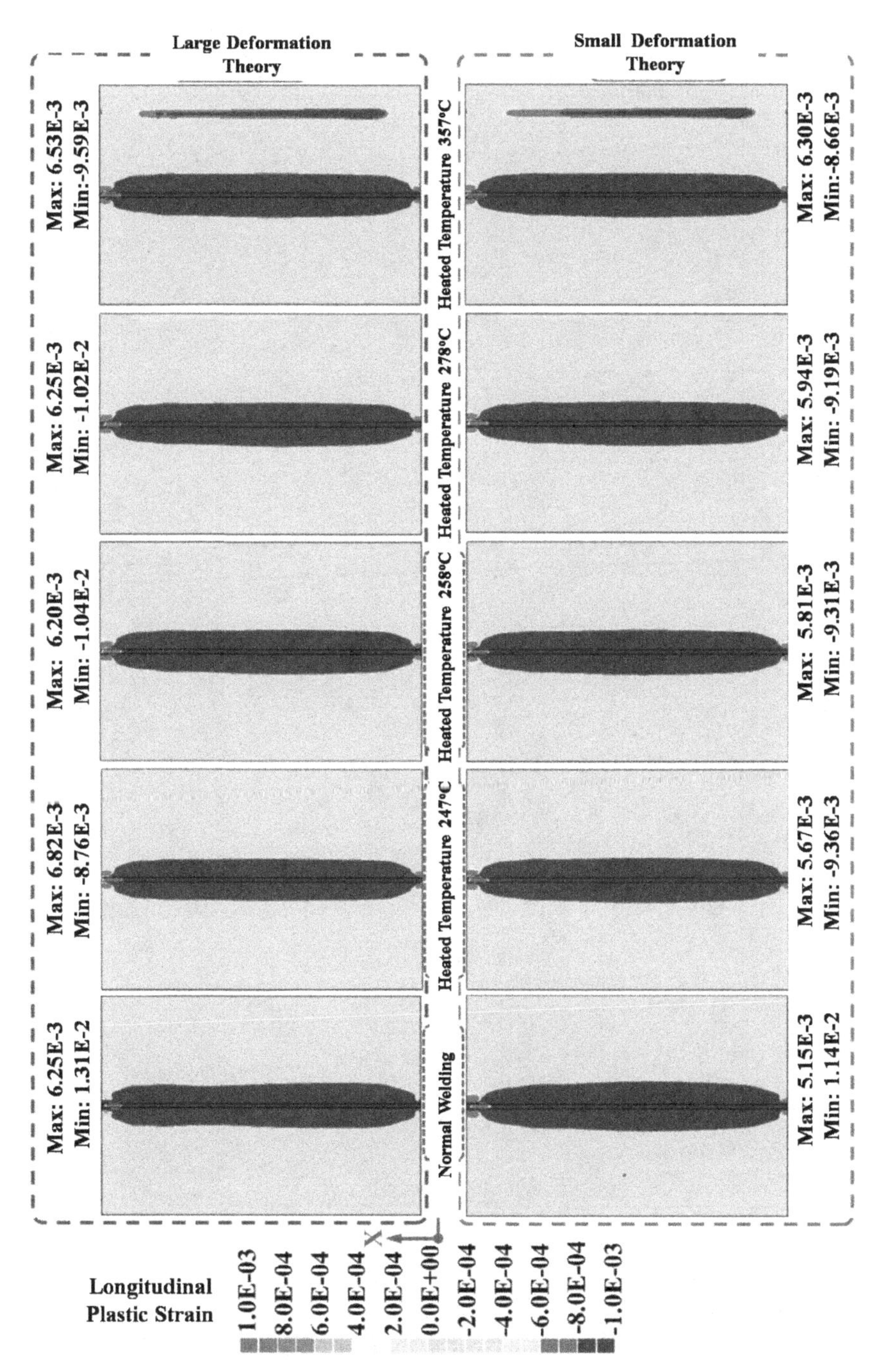

FIGURE 4.9 Contour of plastic strain of examined butt-welded joint by FE computation

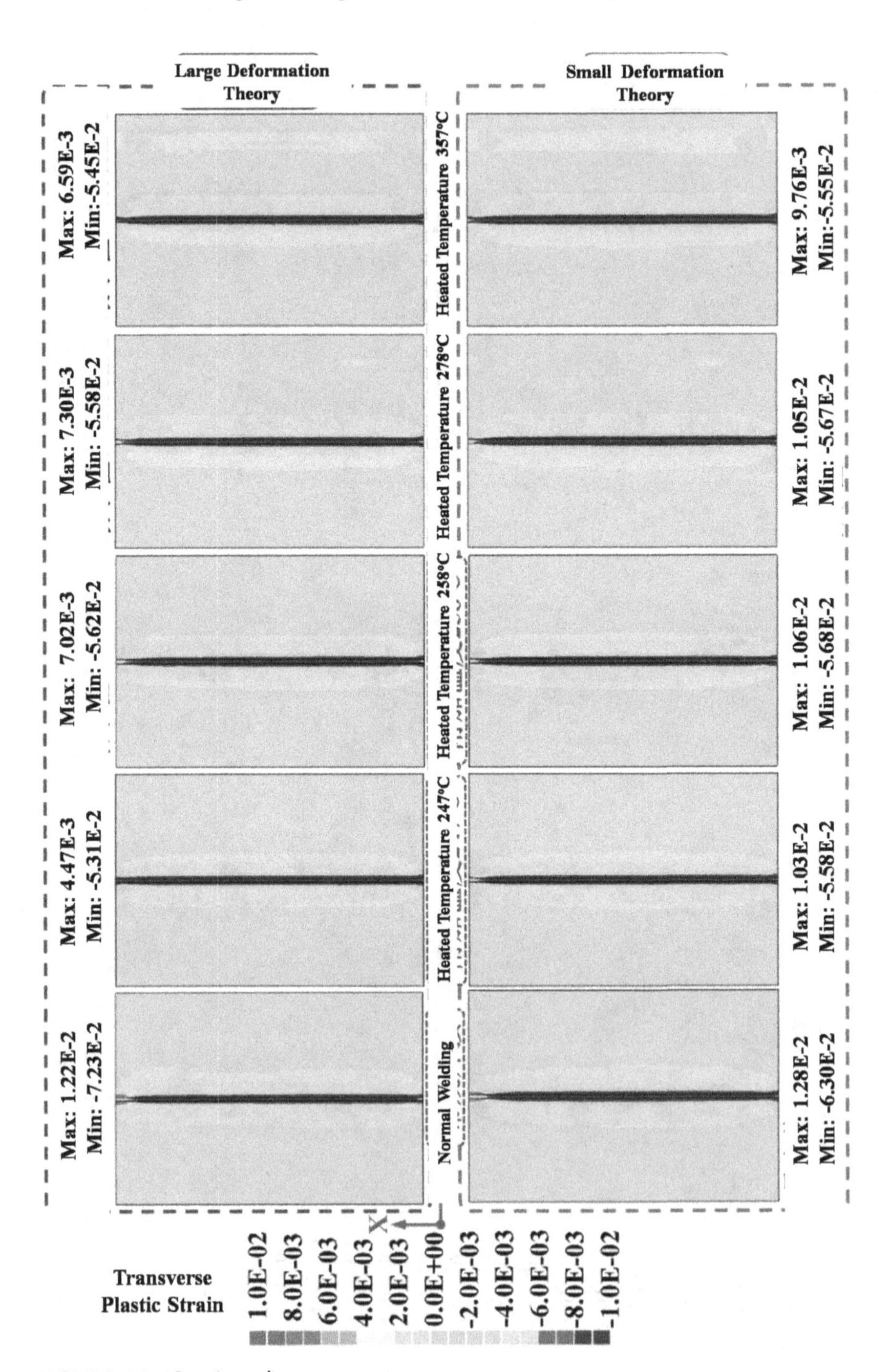

FIGURE 4.9 (Continued)

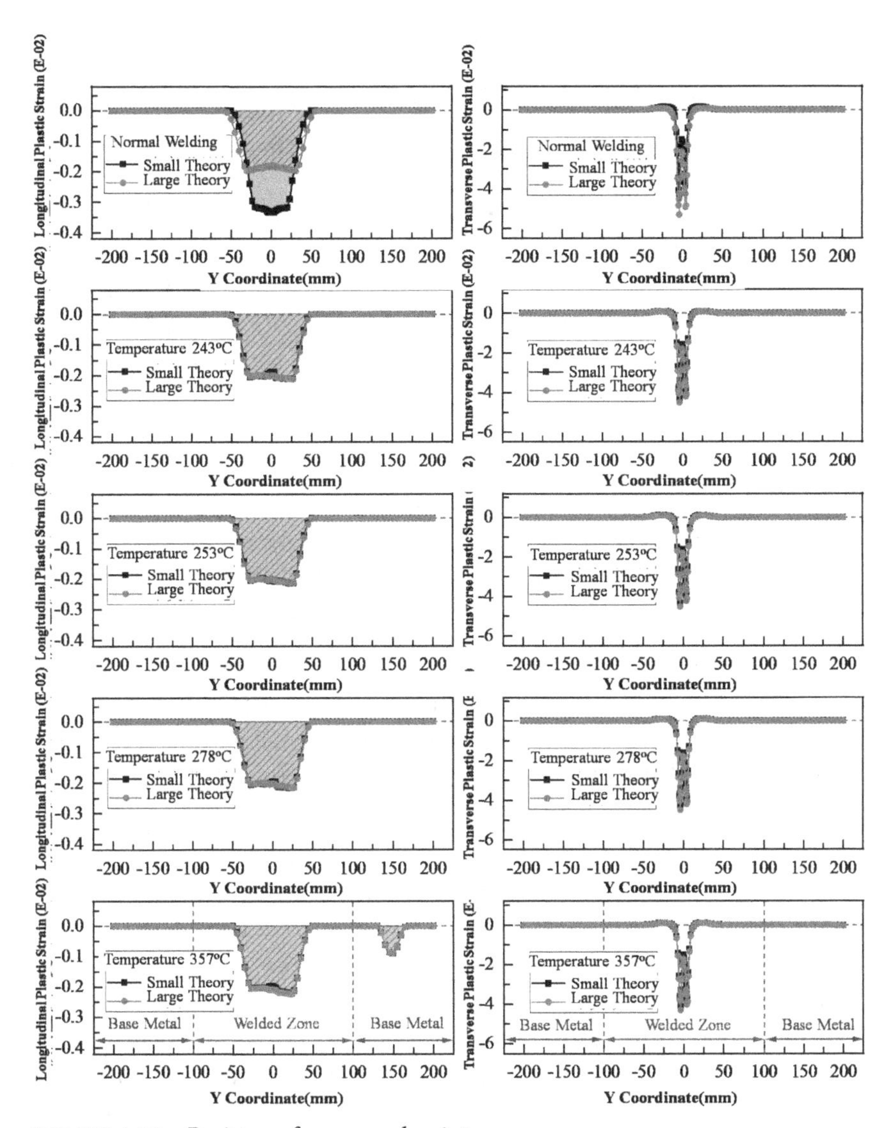

FIGURE 4.10 Position of compared point

buckling is usually considered to be stability problem with geometrical nonlinearity theory.

In conclusion, with the consideration of time consumption, it is preferable to employ small deformation theory to evaluate the distribution and magnitude of welding plastics strain, and large deformation theory should be employed to predict out-of-plane welding distortion as well as buckling deformation during thin plate fabrication.

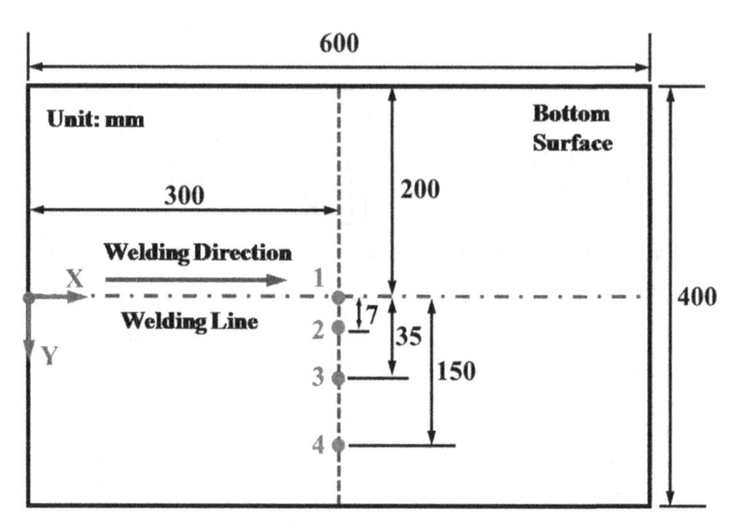

FIGURE 4.11 Distribution of computed longitudinal and transverse plastic strains of points on middle cross-section with FE computation

4.4.2 Evolution of Thermal and Mechanical Behavior

In order to well understand the evolution process of strain, deformation as well as stress due to buckling behavior with normal welding, computed results of butt-welded joint by means of thermal elastic plastic FE computation by considering temperature-dependent material properties and large deformation theory were examined.

As indicated in Figure 4.10, there are four points on the top surface of examined butt-welded joint with AH36 thin plate to be considered. In addition, point 1 is located at the center of welding line (x = 300 mm, y = 0 mm), point 2 is located at the position near the welding line (x = 300 mm, y = 7 mm), point 3 is located at the position near the edge of plastic strain zone (x = 300 mm, y = 35 mm), and the point 4 is located at the region of base material far away the welding line (x = 300 mm, y = 150 mm). The comparison of computed temperature, longitudinal plastic strain, out-of-plane welding distortion, and longitudinal stress of abovementioned four points is demonstrated against time as shown in Figure 4.12

At the beginning of welding, there is no longitudinal plastic strain, out-of-plane welding distortion, or longitudinal stress of considered points, and their temperature is still room temperature due to the large distance between welding arc and considered points. With the movement of welding arc, the temperature of points 1 and 2 will rapidly increase, while

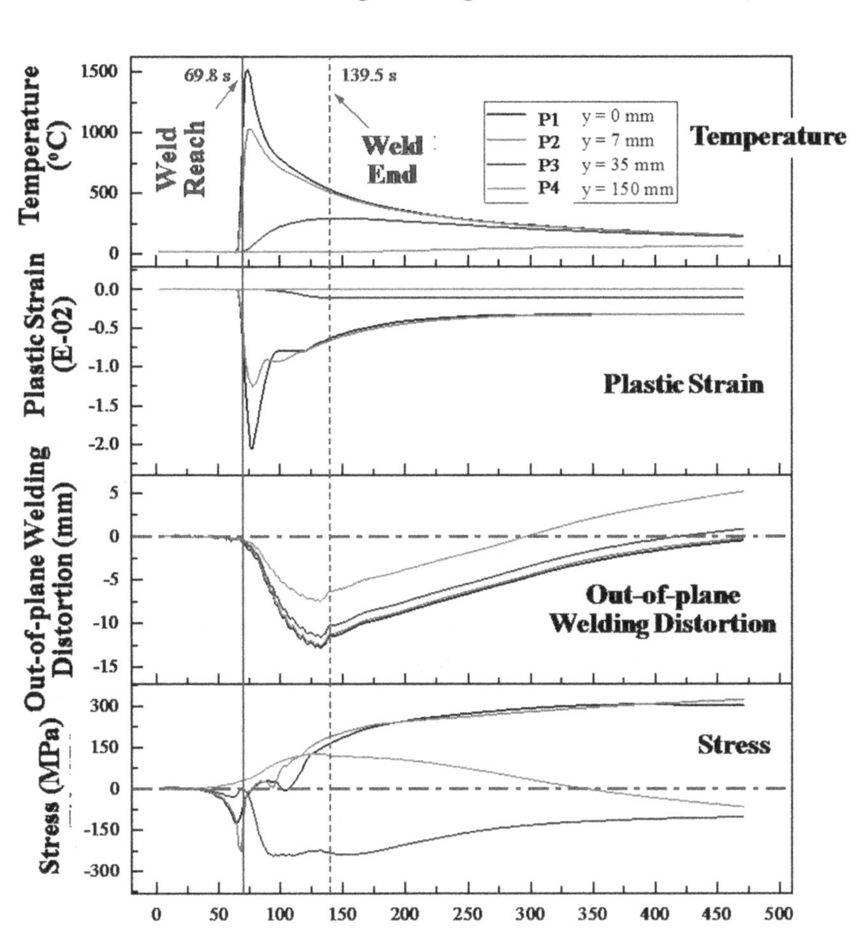

FIGURE 4.12 Variation curve of temperature, plastic strain, out-of-plane welding distortion, and residual stress during conventional welding process

longitudinal compressive plastic strain and longitudinal compressive stress will be generated due to the self-constraint by surrounding base material.

Moreover, the longitudinal compressive plastic strain will reach the maximal magnitude when the transient temperature reaches the highest temperature. When the welding arc passes, the temperature of points 1 and 2 will rapidly decrease, and longitudinal tensile plastic strain will be generated due to the constraint effect of base material, which will offset partial longitudinal compressive plastic strain generated during heating process.

Meanwhile, longitudinal compressive stress will decrease and be converted to be longitudinal tensile stress. With continuous cooling, longitudinal compressive plastic strain of points 1 and 2 will furthermore decrease and keep a constant magnitude, while longitudinal tensile stress will continually increase until the magnitude of yield strength of AH36. Thus, there will be residual longitudinal compressive plastic strain and longitudinal tensile stress in the region of welding line.

For the considered point 3 with lower maximal temperature, longitudinal compressive plastic strain was generated during the heating process, and there is no generation of longitudinal tensile plastic strain during cooling process. With the internal stress equilibrium, the longitudinal compressive stress of point 3 will be generated to balance the longitudinal tensile stress in the region of welding line.

For point 4 in the region of base material far away from the welding line, its maximal temperature could not generate enough thermal strain and thermal stress to yield the base material and obtain longitudinal compressive plastic strain during heating process. Similar to point 3, longitudinal tensile stress will be generated during heating process and longitudinal compressive stress will be then generated during cooling process to balance the longitudinal stress in the region of welding line.

Moreover, it can be seen that deformed directions of out-of-plane welding distortion are different during the heating and cooling processes, while the magnitude of transient welding buckling could reach about 13 mm. This behavior also can confirm the occurrence of welding buckling and its unstable feature.

4.4.3 Comparison of Welding Inherent Deformation

Due to the complicated distribution of welding plastic strain on each cross-section, welding inherent deformation could be evaluated and employed with the integration of all welding plastic strain on same cross-section.

For the computational cases of normal welding and thermal tension techniques with different maximal temperatures in the heated region (247°C, 258°C, 278°C), there is residual plastic strain only in the region near the welding line. Based on the definition, welding inherent deformation can be evaluated for each cross-section.

Meanwhile, when the maximal temperature in the heated region is 357°C, there will be additional plastic strain in the heated region with induction heating. Thus, they will be divided into three regions to individually evaluate inherent deformation for the butt-welded joint of thermal tension techniques with maximal temperature in the heated region (357°C), which are welding line region (-100 mm $\leq y \leq$ 100 mm) and base material region (-200 mm $\leq y < -100$ mm and 100 mm $< y \leq$ 200 mm).

As shown in Figure 4.13, inherent deformation of computational cases with normal welding and different maximal temperature in the heated region by induction heating were all evaluated and demonstrated along the welding line direction. It can be seen that distribution and magnitude of evaluated inherent deformation are almost identical based on computed results with small and large deformation theory, and the magnitudes of inherent deformation components generated with normal welding are larger than that with transient thermal tension technique. In particular, the magnitudes of longitudinal inherent shrinkage and transverse inherent bending were obviously reduced compared to that with normal welding.

When the end effect (position of welding beginning and completion) is ignored, there will be a uniform distribution of inherent deformation in the middle region. Thus, the region (100–500 mm) with uniform distribution was then examined to evaluate the certain value of inherent deformation for examined buttwelded joint with different welding procedures, which is summarized in Table 4.4.

In addition, the minus inherent deformation means shrinkage deformation and the minus inherent bending means downward bending deformation. The evaluated magnitudes of inherent deformation of examined butt-welded joint with AH36 thin plate with transient thermal tension technique based on computed results of thermal elastic plastic FE computation with small and large deformation theory are almost identical. However, the evaluated magnitudes of inherent deformation with normal welding based on computed results of thermal elastic plastic FE computation with small and large deformation theory are a little bit different, while the differences of evaluated magnitudes of longitudinal inherent shrinkage and transverse inherent bending are 11.4% and 7.3%, respectively.

It can be understood that due to serious welding buckling behavior with normal welding, thermal elastic plastic FE computation of examined butt-welded joint with AH36 thin plate by means of small deformation theory could not accurately predict the out-of-plane welding distortion as well as buckling deformation, and its evaluated inherent deformation definitely has difference comparing with that based on FE computation with considering large deformation theory.

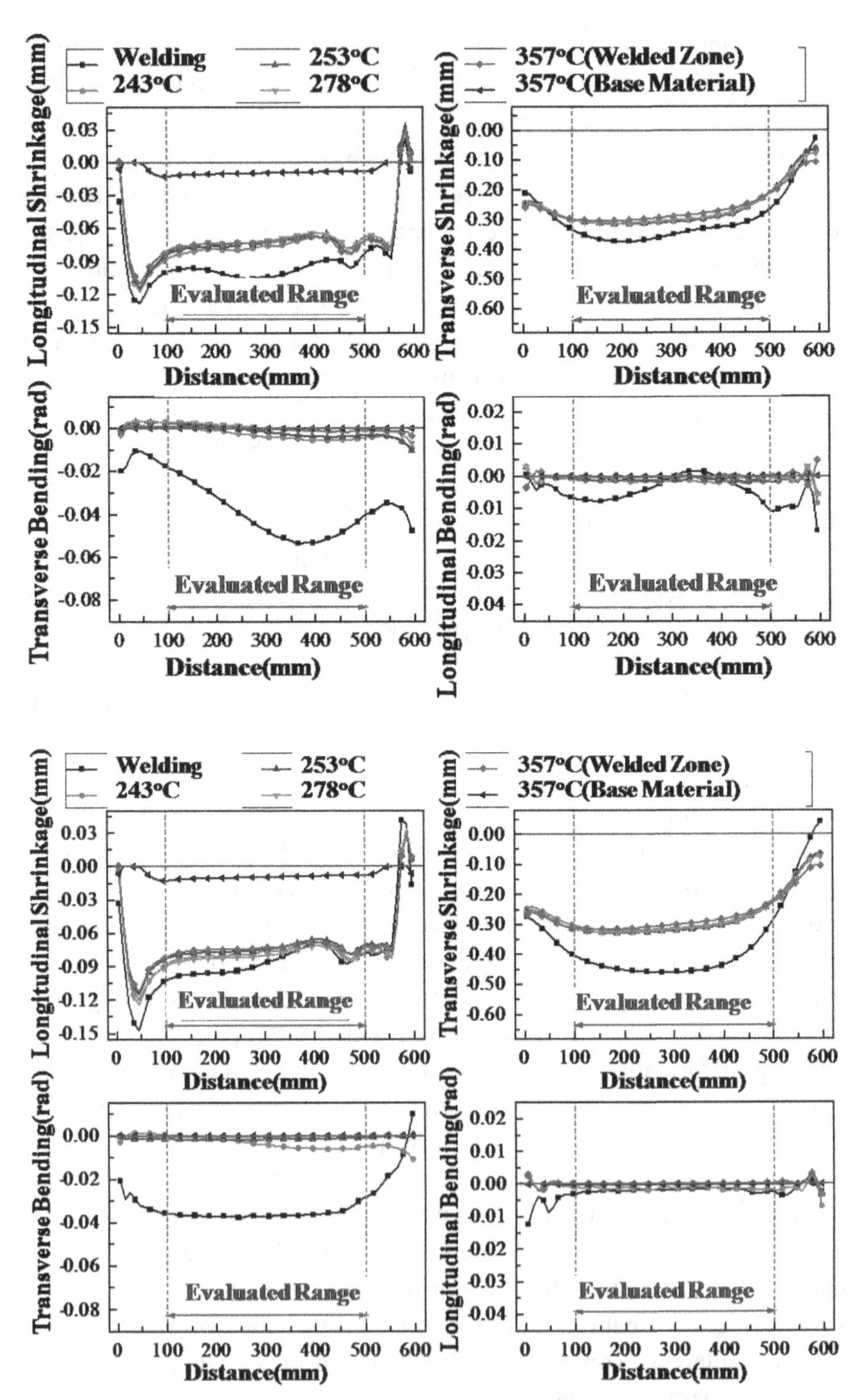

FIGURE 4.13 Evaluated welding inherent deformation with computed results

TABLE 4.4 Magnitude of welding inherent deformation evaluated by computed results of thermal elastic plastic FE analysis.

		Longitudinal Inherent Shrinkage (mm)	Transverse Inherent Shrinkage (mm)	Transverse Inherent Bending (rad)	Longitudinal Inherent Bending (rad)
Conventional Welding	Large Deformation	−0.0976	−0.3422	−0.0411	−0.0038
	Small Deformation	−0.0865	−0.4302	−0.0381	−0.0025
Thermal Tension with Temperature of 243°C	Large Deformation	−0.0768	−0.2955	−0.0033	−0.0012
	Small Deformation	−0.0770	−0.3008	−0.0036	−0.0014
Thermal Tension with Temperature of 253°C	Large Deformation	−0.0735	−0.2948	−0.0013	−0.0011
	Small Deformation	−0.0738	−0.3086	−0.0016	−0.0005
Thermal Tension with Temperature of 278°C	Large Deformation	−0.0715	−0.2919	−0.0005	−0.0003
	Small Deformation	−0.0725	−0.3059	−0.0004	−0.0004
Thermal Tension with Temperature of 357°C (region near welding line)	Large Deformation	−0.0735	−0.2822	−0.0003	−0.0002
	Small Deformation	−0.0734	−0.2938	−0.0003	−0.0002
Thermal Tension with Temperature of 357°C (region of base material)	Large Deformation	−0.0097	−0.0001	0.0000	0.0000
	Small Deformation	−0.0095	−0.0002	0.0000	0.0000

With the transient thermal tension technique, welding-induced buckling can be effectively reduced and welding buckling behavior has less influence on welding inherent deformation. Thus, there will be almost identical magnitudes of welding inherent deformation of examined butt-welded joint with AH36 thin plate evaluated by computed results with considering small deformation and large deformation theory.

Moreover, the magnitude of transverse inherent shrinkage is much larger than that of longitudinal inherent shrinkage due to their different self-constraint supported by surrounding base material during welding

[67, 80]. The lowest magnitude of longitudinal inherent shrinkage in the region near welding line can be obtained with the transient thermal tension of maximal temperature (278°C) in the heated region. Compared with normal welding, transient thermal tension technique can maximally reduce longitudinal inherent shrinkage by 26.7% and transverse inherent bending by 90%.

With the mechanism investigation of welding buckling occurrence, in-plane shrinkage, in particular, longitudinal shrinkage force, is the dominant cause of welding buckling of butt-welded joint, and reduction of longitudinal shrinkage force will effectively mitigate welding buckling or avoid the occurrence of welding buckling. Meanwhile, welding inherent bending will be generated due to the nonuniform distribution of residual plastic strain in the thickness direction, which will be the disturbance to trigger the occurrence of welding buckling and influence the magnitude of out-of-plane welding distortion. Compared with transverse inherent bending, longitudinal inherent bending also considered disturbance to trigger the occurrence of welding buckling is usually ignored due to its lesser magnitude.

4.4.4 Variation of Welding Inherent Deformation

Mitigation mechanism of welding-induced buckling with transient thermal tension technique was clarified with variation of longitudinal plastic strain during the heating and cooling processes. Taking the computed results of thermal elastic plastic FE computation with considering both temperature-dependent material properties and large deformation with transient thermal tension (maximal temperature of the heated region: 278°C; transverse distance d1: 150 mm and longitudinal distance d2: 0 mm) for further investigation, variation of longitudinal plastic strains of Points A and B located at the center position of top and bottom surfaces of welding line were demonstrated in Figure 4.14.

It can be seen that longitudinal compressive plastic strains of Points A and B will be generated and increased with the temperature elevation during the heating process, while longitudinal tensile plastic strains will be produced during the cooling process to offset partial longitudinal compressive plastic strain. Moreover, the maximal temperature of welded zone will not be influenced by induction heating, less magnitude of longitudinal compressive plastic strains of Points A and B will be generated during the heating process with the transient thermal tension, and after cooling down

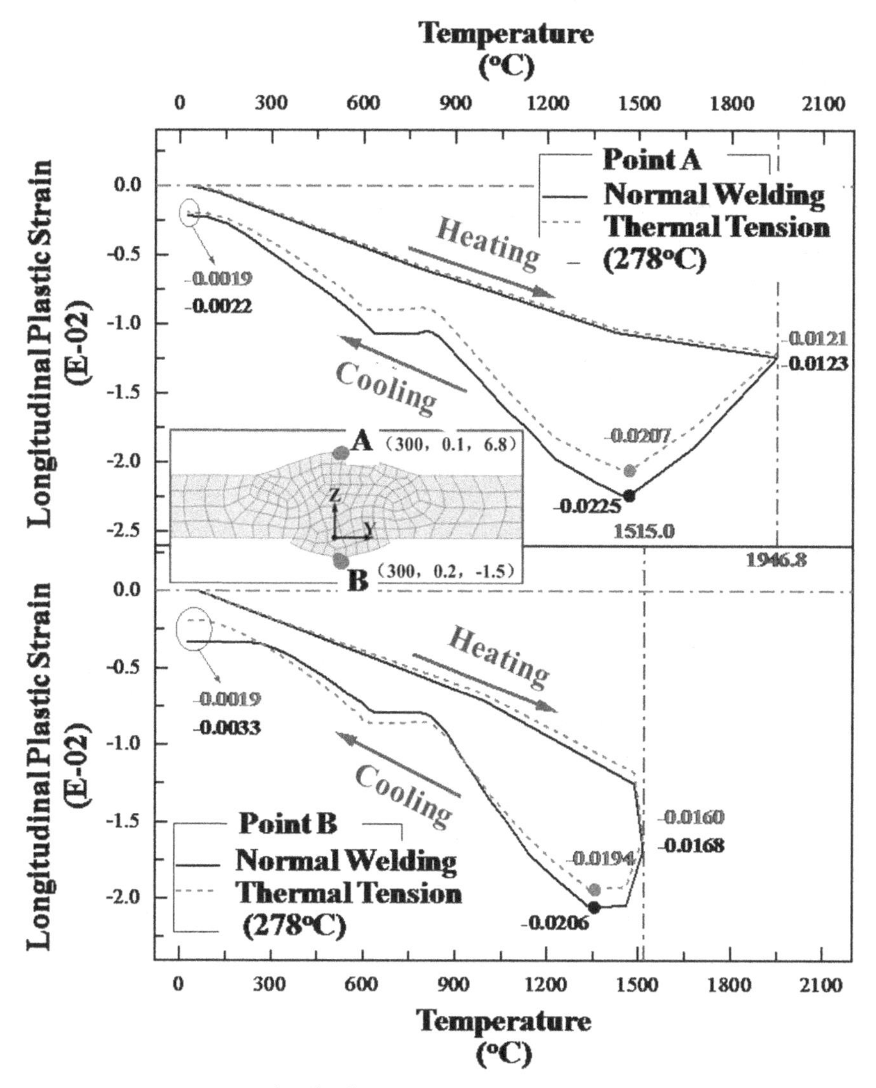

FIGURE 4.14 Longitudinal plastic strain against temperature with normal welding and transient thermal tension

to room temperature, residual longitudinal compressive plastic strains with the transient thermal tension is less than that with normal welding.

The magnitudes of longitudinal plastic strain are then summarized in Table 4.5. It can be seen that the variation tendency of longitudinal plastic strain is identical to that demonstrated in Figure 4.14, while less magnitude of longitudinal compressive plastic strain is generated during the

TABLE 4.5 Longitudinal plastic strains with normal welding and transient thermal tension.

Welding Procedure	Point A Normal Welding	Point A Transient Thermal Tension	Point B Normal Welding	Point B Transient Thermal Tension
Longitudinal compressive plastic strain during heating	-0.0123	-0.0121	-0.0168	-0.0160
Longitudinal tension plastic strain during cooling	0.0101	0.0102	0.0135	0.0141
Residual longitudinal compressive plastic strain	-0.0022	-0.0019	-0.0033	-0.0019

heating process and more magnitude of longitudinal tensile plastic strain is generated during the cooling process. Thus, magnitude of residual longitudinal compressive plastic strain will be eventually reduced.

Due to the nature of shrinkage in the longitudinal direction along the welding line, longitudinal inherent shrinkage deformation is usually converted to be longitudinal inherent shrinkage force as well as tendon force [65, 67]. With the definition of welding inherent deformation, residual longitudinal compressive plastic strain on each cross section can be employed to evaluate the longitudinal shrinkage force, while only normal welding and transient thermal tension (maximal temperature of the heated region: 278°C; transverse distance d1: 150 mm and longitudinal distance d2: 0 mm) were considered.

Distribution of longitudinal shrinkage force along the welding line was obtained and demonstrated in Figure 4.15. In addition, longitudinal shrinkage force with transient thermal tension of examined butt-welded joint of AH36 thin plate was reduced from −205.1 KN with normal welding to −150.9 KN by about 26.4%. Therefore, transient thermal tension will reduce the magnitude of residual longitudinal compressive plastic strain and longitudinal inherent shrinkage force to mitigate the welding buckling deformation.

Meanwhile, transverse inherent bending as disturbance will trigger the occurrence of welding buckling and influence the magnitude of

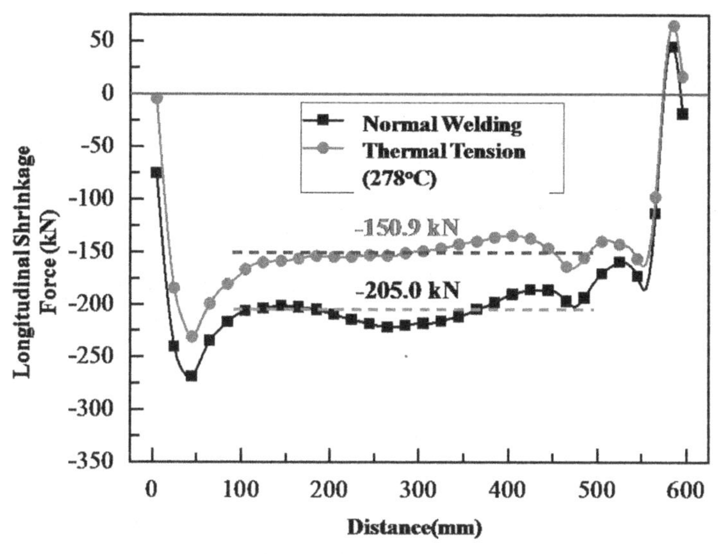

FIGURE 4.15 Comparison of longitudinal inherent shrinkage force with normal welding and transient thermal tension

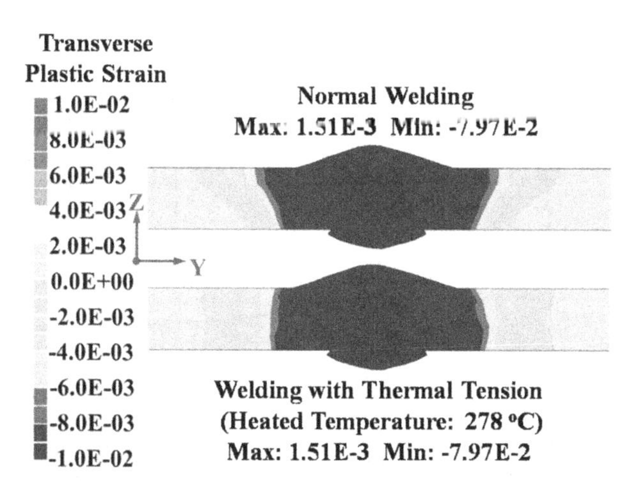

FIGURE 4.16 Comparison of transverse plastic strain on middle cross-section with normal welding and transient thermal tension

out-of-plane welding distortion. As shown in Figure 4.16, contour plotting of residual transverse plastic strain on middle cross-section of butt-welded joint with normal welding and transient thermal tension (maximal temperature of the heated region: 278°C) were compared.

For normal welding, the region with residual transverse plastic strain near the top surface is larger than that near the bottom surface, and transverse inherent bending as well as downward out-of-plane welding distortion will then be generated. With the transient thermal tension, residual transverse plastic strain has uniform distribution in the thickness direction and less magnitude of transverse inherent bending was then generated.

Moreover, the magnitude of transverse inherent bending will be determined not only by temperature gradient in the thickness direction but also by the transient behavior of buckling deformation when welding buckling occurs [94]. Thermal cycles of Points C and C′ with normal welding and transient thermal tension were compared as demonstrated in Figure 4.17.

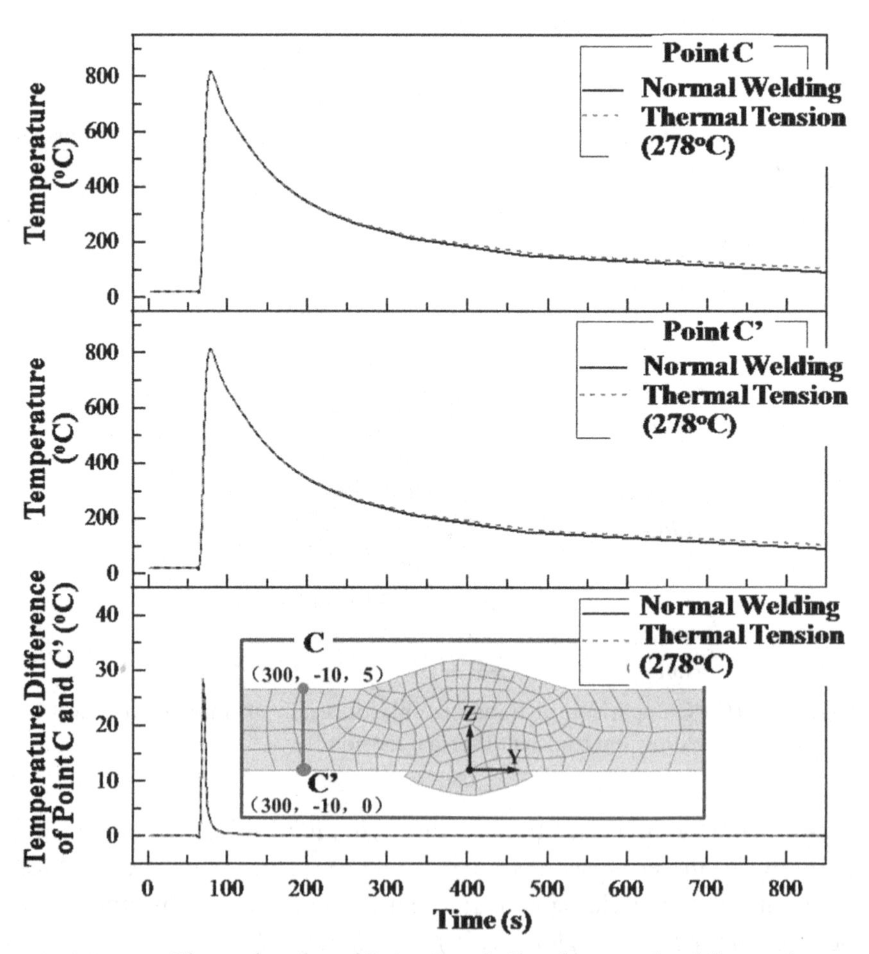

FIGURE 4.17 Thermal cycles of Point C and C′ with normal welding and transient thermal tension

It can be seen that transient thermal tension with induction heating will not influence the thermal cycles of points C and C′ in the region of welding line, and temperature difference between points C and C′ for the case with transient thermal tension is almost constant. Thus, variation of transverse inherent bending with transient thermal tension resulted from the transient behavior of welding buckling.

When the welding buckling occurs, transverse plastic strain will be determined by transient out-of-plane welding distortion with large magnitude, while the temperature gradient in the thickness direction is not the unique governing cause. As shown in Figure 4.18, out-of-plane welding distortion of point C against time was observed. It can be seen that out-of-plane welding distortion of point C is bending downward at –12 mm firstly and then bending upward with eventual magnitude of 2.3 mm due to the serious welding buckling behavior with normal welding. With the transient thermal tension, variation of out-of-plane welding distortion of point C is effectively mitigated compared to that with normal welding.

In conclusion, transient thermal tension can effectively mitigate the welding-induced bucking not only by reduction of longitudinal shrinkage force but also by reduction of transient inherent bending due to more uniform distribution of residual transverse plastic strain in the thickness direction. Less disturbance of transient inherent bending is also beneficial to mitigate the welding-induced bucking.

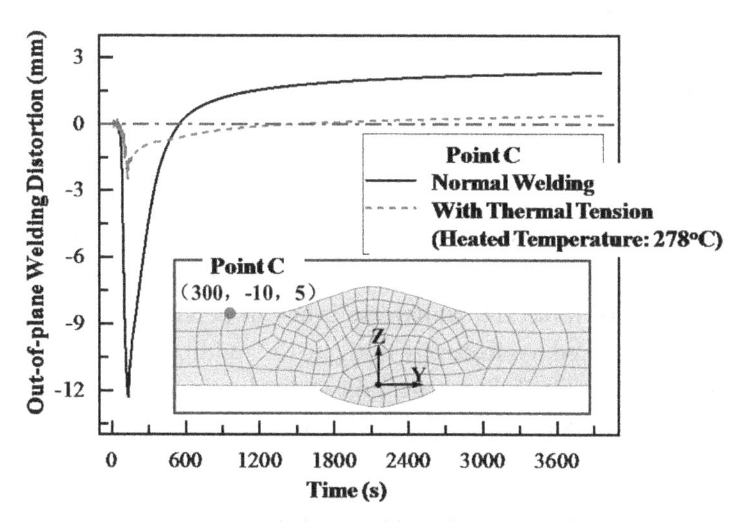

FIGURE 4.18 Transient out-of-plane welding distortion of point C with normal welding and transient thermal tension

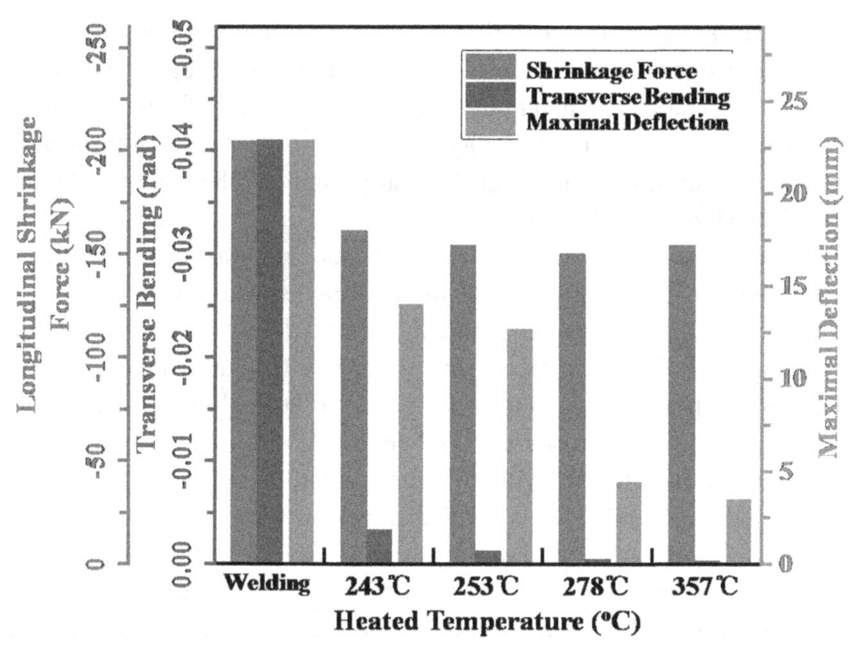

FIGURE 4.19 Tendency of longitudinal shrinkage force, transverse inherent bending, and maximal relative out-of-plane welding distortion of the region of welding line

Tendency of longitudinal shrinkage force, transverse inherent bending, and maximal relative out-of-plane welding distortion of the region of welding line with normal welding and transient thermal tension was then demonstrated in Figure 4.19. Longitudinal shrinkage force as well as transverse inherent bending with transient thermal tension will both be reduced, while their variation tendency has agreement compared to maximal relative out-of-plane welding distortion. Thus, with the above investigation, transient thermal tension with induction heating will have thermal tension effect on welding line and reduce the constraint intensity of welding line supported by surrounding base material, while the magnitudes of longitudinal shrinkage force and transverse inherent bending during welding process were reduced to effectively mitigate the welding induced buckling.

In detail, longitudinal shrinkage force is the dominant cause to generate welding buckling, and transverse inherent bending as disturbance will trigger the occurrence of welding buckling and influence the magnitude of out-of-plane welding distortion. Reduction of both longitudinal shrinkage force and transverse inherent bending will be preferred for mitigation of welding-induced buckling.

4.5 SUMMARY

Transient thermal elastic plastic FE computation was carried out to represent the thermal and mechanical behavior during thin plate butt welding with solid elements model. Computed thermal cycles have a good agreement compared with measurement, and predicted out-of-plane welding distortion also has good agreement compared with measurement not only for tendency but also for magnitude of welding distortion, when the large deformation theory was employed. Welding inherent strain and inherent deformation were then examined to clarify the mechanism of transient thermal tension to mitigate the welding-induced bucking, while longitudinal plastic strain and longitudinal shrinkage force as the elementary cause were reduced.

CHAPTER 5

Parameter Sensitivity Analysis of Thermal Tension Techniques

WITH THE ABOVE INVESTIGATION, the mechanism of transient thermal tension to mitigate welding-induced buckling can be understood with the reduction of longitudinal shrinkage force as well as tendon force.

Longitudinal shrinkage force is mainly determined by residual compressive plastic strain, so-called welding inherent strain, which can be obtained by the difference of compressive plastic strain generated during heating and tensile plastic strain generated during cooling. With the transient thermal elastic plastic FE computation with ISM (iterative substructure method) and parallel computation, a series of FE analyses were carried out to examine the parameter sensitivity of thermal tension techniques, such as intensity of induction heating, and transverse distance between induction heating coil and welding line.

5.1 INFLUENCE OF INDUCTION HEATING INTENSITY

During the engineering application of transient thermal tension, maximal temperature of heated region with induction heating, which is easily understood and controlled, was considered to replace the magnitude of induction heating.

 DOI: 10.1201/9781003442523-5

Taking the butt-welded joint of BT-3 with the transient thermal tension as benchmark, influence research of maximal temperature of heated region with induction heating on welding bucking mitigation was then carried out. Similar to actual experiment with application of transient thermal tension, there is only one heat source of induction heating, which is located at one side of welding line of butt-welded joint.

In addition, the transverse distance (d1) between centers of induction heating and welding line is 150 mm, as well as the longitudinal distance (d2) equals 0 mm; the induction heating will move with welding arc simultaneously. The shape parameters of body heat source model validated above as demonstrated in the FE computation of butt-welded joint of BT-3 were employed again, while the energy density of body heat source will be considered to examine the influence of maximal temperature of heated region with induction heating.

Based on the solid elements FE model of butt-welded joint as shown in Figure 4.1, thermal analysis with four considered maximal temperatures of heated region with induction heating (196°C, 258°C, 278°C, and 357°C) was carried out to examine the temperature profile, while computed transient temperature contour plotting was illustrated in Figure 5.1.

Different from conventional welding, both welding region and induction heating region of butt-welded joint with transient thermal tension have temperature variation, while their magnitudes of maximal temperature have large differences. It also can be seen that the induction heating will only influence the heated region of base material and not influence the temperature profile of the region near welding line.

Taking the computed temperature filed as thermal loading to solid elements FE model, mechanical analysis considering the material and geometry nonlinearity was then carried out to evaluate the out-of-plane welding distortion of butt-welded joints with different magnitudes of induction heating as demonstrated in Figure 5.2. In addition, the buckling feature of butt-welded joint was still observed with the upward and downward out-of-plane welding distortions when the transient thermal tension was considered; however, the magnitude of out-of-plane welding distortion was significantly reduced.

Furthermore, in order to consider the influence of transient thermal tension with different magnitudes of induction heating on out-of-plane welding distortion, maximal relative out-of-plane welding distortion as

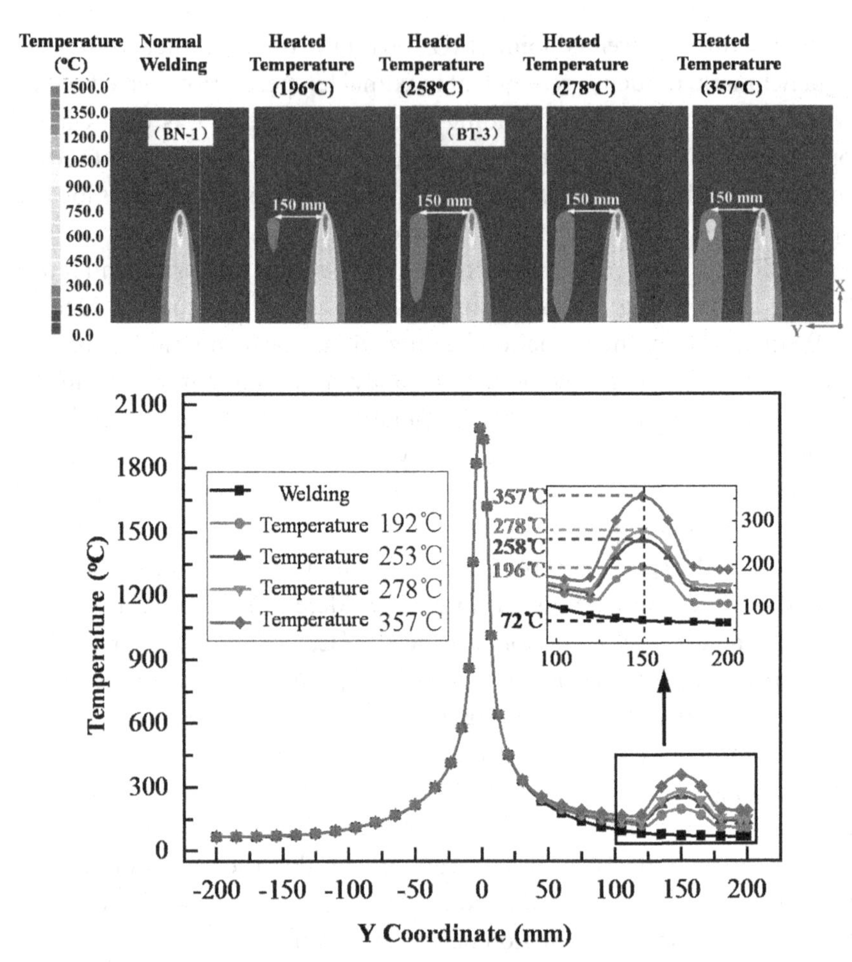

FIGURE 5.1 Temperature distribution of butt welding with different magnitudes of induction heating

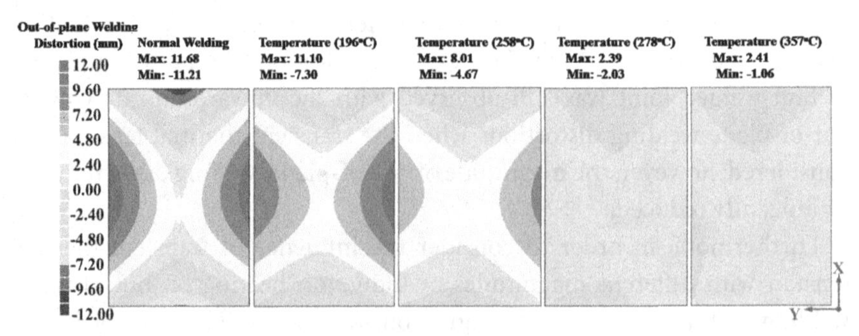

FIGURE 5.2 Out-of-plane welding distortion distribution of butt welding with different magnitudes of induction heating

defined by Eq. (5.1) as well as overall standard deviation of butt-welded joint as defined by Eq. (5.2) were proposed as demonstrated in Figure 5.2.

$$U_{max} = Z_{max} - Z_{min} \tag{5.1}$$

$$\sigma = \sqrt{\frac{1}{n}\sum_{i=1}^{n}(x_i - \bar{x})^2} \tag{5.2}$$

where U_{max} is the maximal relative out-of-plane welding distortion, mm; Z_{max} is the maximal out-of-plane welding distortion, mm; Z_{min} is the minimal out-of-plane welding distortion, mm; σ means the overall standard deviation, mm; x_i is the out-of-plane welding distortion of points on upper surface of butt-welded joint, mm; and n is the total number of points on upper surface.

It can be seen that reduction effect of out-of-plane welding distortion will be obvious with the increasing magnitude of induction heating. In addition, the maximal temperature of heated region with induction heating equals 357°C with best reduction effect of out-of-plane welding distortion; maximal relative out-of-plane welding distortion and overall standard deviation can be reduced by 84.8% and 85.8%, respectively.

When maximal temperature of heated region with induction heating equals 278°C, maximal relative out-of-plane welding distortion and overall standard deviation can be reduced by 80.6% and 83.4%, respectively. However, maximal relative out-of-plane welding distortion and overall standard deviation can only be reduced by 19.6% and 21.1%, respectively, When maximal temperature of heated region with induction heating equals 196°C. Thus, the reduction effect of out-of-plane welding distortion by means of transient thermal tension is not obvious when the maximal temperature of heated region with induction heating is lower.

The out-of-plane welding distortion can be well reduced by transient thermal tension when the maximal temperature of heated region with induction heating reaches a certain value. However, the maximal temperature of heated region with induction heating could not be too high, and additional plastic strain due to induction heating will be generated, which will be harmful for out-of-plane welding distortion mitigation. Therefore, there will be an appropriate range of maximal temperature of heated region

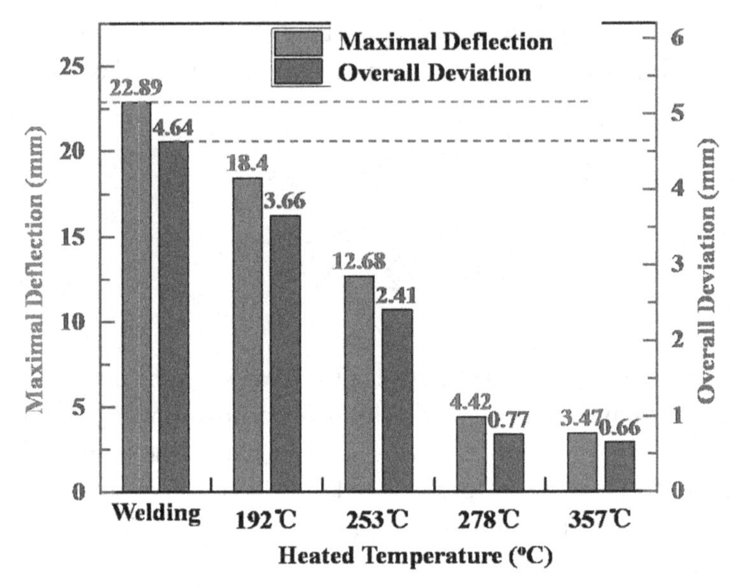

FIGURE 5.3 Comparison of out-of-plane welding distortion of butt-welded joints with different magnitudes of induction heating

with induction heating, and out-of-plane welding distortion of examined butt-welded joint can be effectively reduced to enhance the accuracy of fabrication.

As shown in Figure 5.3, the distribution and magnitude of out-of-plane welding distortion are both similar when the maximal temperatures of heated region with induction heating are 278°C and 357°C. In addition, the lower maximal temperatures of heated region with induction heating at 278°C will be selected during the actual fabrication due to the lower cost consumption and higher economic benefits.

As shown in Figure 5.4, longitudinal plastic strains along the welding line of examined butt-welded joints with conventional welding and welding with transient thermal tension with different magnitudes of induction heating can be observed. There is a compressive plastic strain in the region of welding line with the conventional welding, and its distribution will be uniform in the middle region along the welding line.

With the transient thermal tension, there will be compressive plastic strain in the region of welding line as demonstrated in Figure 6.3 with the maximal temperatures of 196°C, 258°C, and 278°C, while maximal temperatures of heated region with induction heating are less than the yield temperature. However, when the maximal temperatures of heated

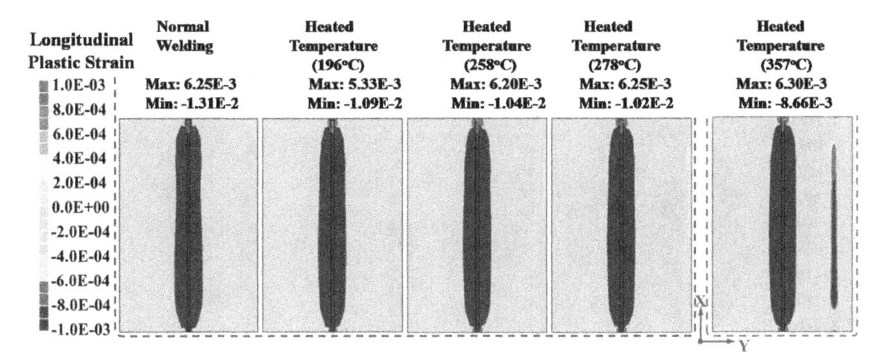

FIGURE 5.4 Distribution of longitudinal plastic strain of butt-welded joints with different magnitudes of induction heating

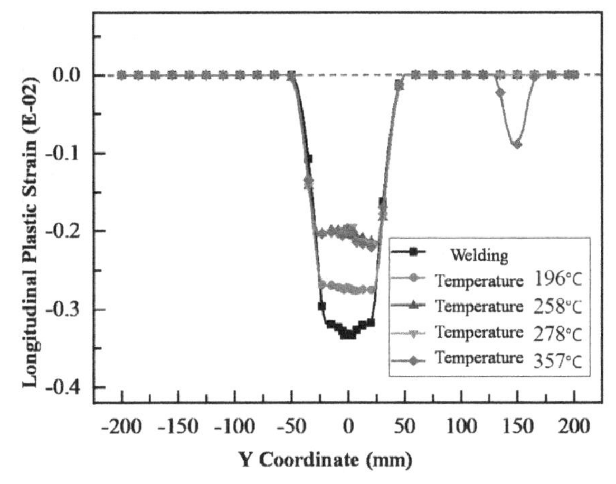

FIGURE 5.5 Comparison of longitudinal plastic strain on middle cross-section of butt-welded joints with different magnitudes of induction heating

region with induction heating reach 357°C, there will be additional compressive plastic strain generated in the region of heated region with induction heating. The distribution of longitudinal plastic strain in the middle cross-section of examined butt-welded joint can also be obtained as shown in Figure 5.5. Transient thermal tension could reduce the magnitude of longitudinal plastic strain in the region of welding line, and the distribution region of longitudinal plastic strain almost locates a constant region with width of 100 mm. There will be additional compressive plastic strain generated in the heated region with induction heating when the maximal temperatures of heated region equal 357°C, and both the region and

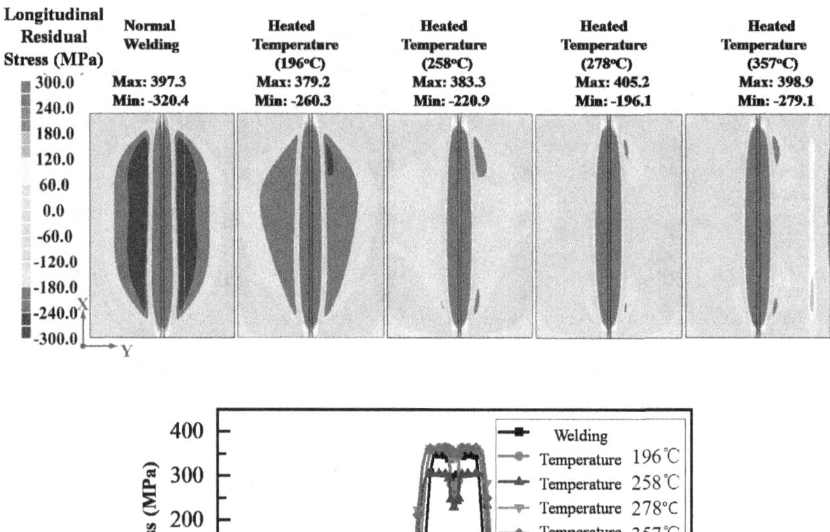

FIGURE 5.6 Comparison of longitudinal residual stress without and with thermal tension with different magnitudes of induction heating

magnitude of additional compressive plastic strain are much less compared with welding plastic strain.

With the above-computed results, there are two cases to reduce the out-of-plane welding distortion, as well as welding buckling of examined butt-welded joint, which are without and with additional compressive plastic strains generated in the region of heated region with induction heating. In actual welding fabrication, the case without additional compressive plastic strain generated in the region of heated region with induction heating will be preferred, which is simple and effective to reduce the welding distortion with less cost consumption. The case with additional compressive plastic strain generated in the region of heated region with induction heating may be useful for welding distortion mitigation after welding.

Moreover, as shown in Figure 5.6, longitudinal residual stresses along the welding line of examined butt-welded joints with conventional welding and welding with transient thermal tension with different magnitudes of induction heating can be observed. It can be seen that maximal compressive residual stress with maximal temperatures of heated region with induction heating of 278°C is only -196.1 MPa, which is reduced by 38.8% compared with -320.4 MPa by conventional welding. With reduction of compressive residual stress, the mechanism of transient thermal tension to mitigate welding buckling of butt-welded joint was also understood.

5.2 INFLUENCE OF DISTANCE BETWEEN COIL AND ARC

Based on the experimental butt welded joint of BT-3, the maximal temperatures of heated region with induction heating were considered to be 258°C, and the induction heating will move simultaneously with the welding arc by identical speed, which means the longitudinal distance d2 between induction heating coil and welding arc is 0 mm.

The transverse distance d1 between induction heating coil and welding arc was considered with different values of 100 mm, 130 mm, 150 mm, and 170 mm to examine its influence on reduction of welding buckling of thin plate butt welding. A series of transient thermal elastic plastic FE computations with different transverse distance d1 of butt-welded joint were carried out, and the computed results of thermal analysis were demonstrated in Figure 5.7. Similar to previous computations, additional heat source due to induction heating will not influence the temperature distribution in the region of welding line compared with that with conventional welding.

Due to the identical magnitude of induction heating, additional temperature profiles are almost identical and maximal temperatures of heated region with induction heating are all 258°C. Meanwhile, the temperature profile in the region of base material has large difference due to the different transverse distance d1. The edge of examined butt-welded joint will have higher temperature with the increasing transverse distance d1.

Computed out-of-plane welding distortions of butt-welded joint with different transverse distance d1 were also obtained as demonstrated in Figure 5.8. The out-of-plane welding distortion will be smallest with the best reduction effect when the transverse distance d1 equals 170 mm.

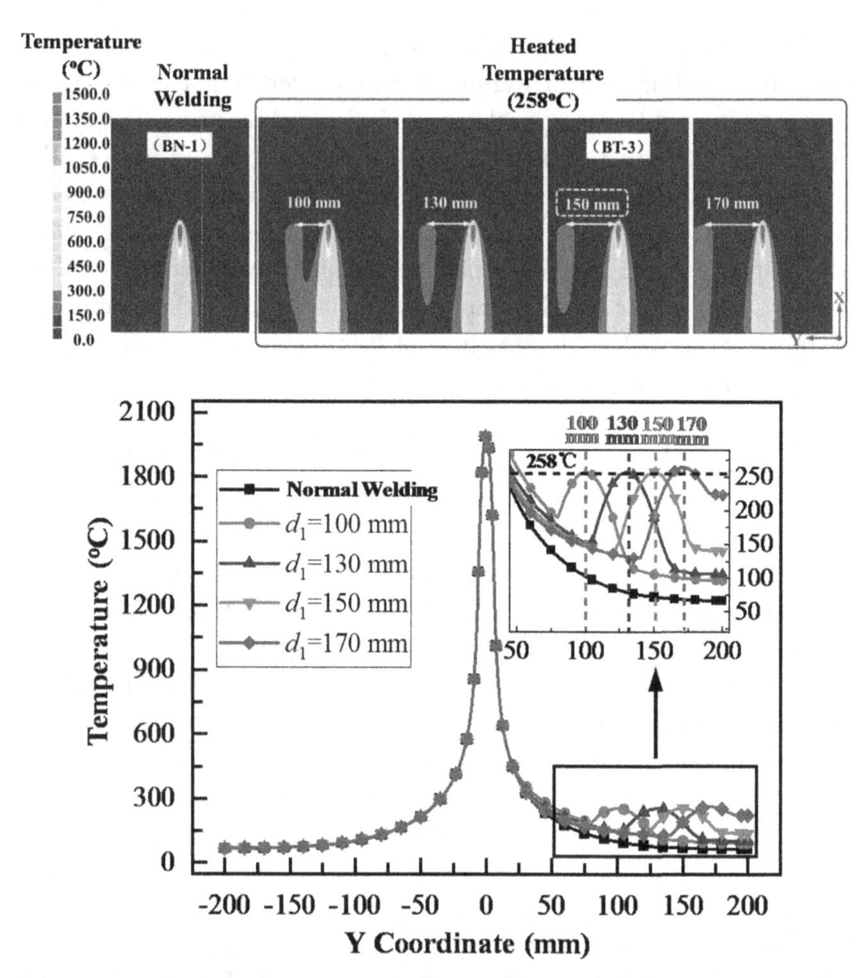

FIGURE 5.7 Computed temperature of butt welding with different transverse distance (d1) of induction heating

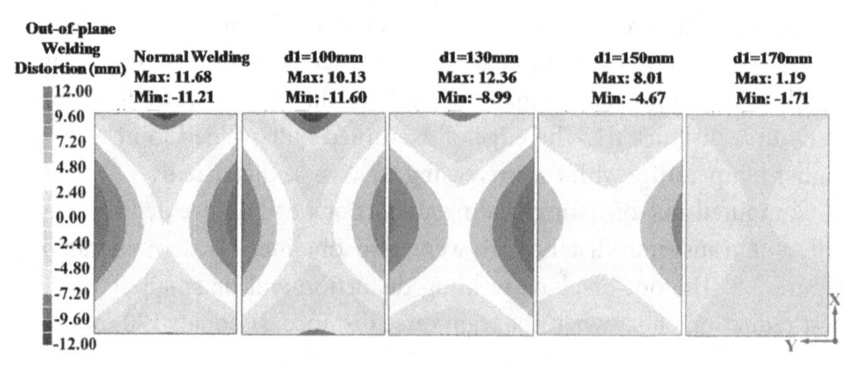

FIGURE 5.8 Out-of-plane welding distortion distribution of butt welding with different transverse distance (d1) of induction heating

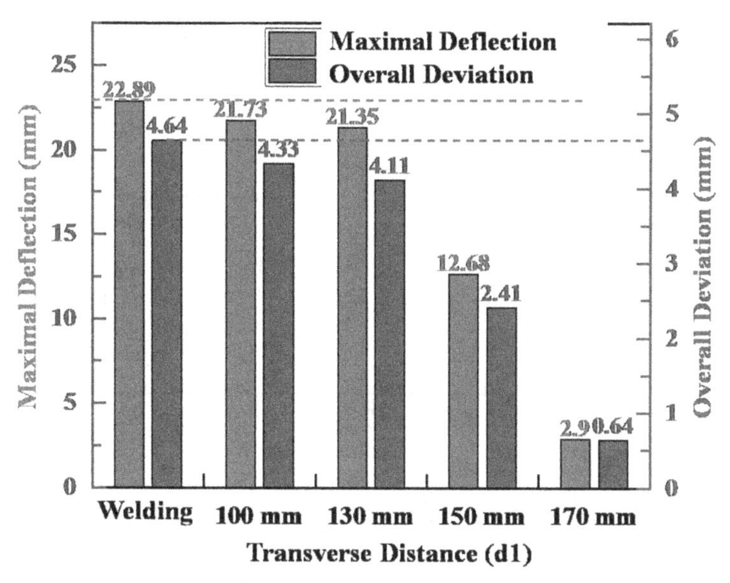

FIGURE 5.9 Comparison of out-of-plane welding distortion of butt-welded joints with different transverse distance (d1) of induction heating

The maximal relative out-of-plane welding distortion and overall standard deviation of butt-welded joint were then shown in Figure 5.9. It can be seen that the out-of-plane welding distortion can be obviously reduced when the transverse distance d1 is 150 mm and 170 mm. In particular, the maximal relative out-of-plane welding distortion and overall standard deviation of butt-welded joint can be reduced by 87.3% and 86.2%, respectively, to enhance the fabrication accuracy, when the transverse distance d1 is 170 mm.

With the FE computation and analysis, the transverse distance d1 should be bigger than 150 mm to effectively reduce the out-of-plane welding distortion for the examined butt-welded joint and applied thermal tension technique. Otherwise, the reduction effect of out-of-plane welding distortion could be ignored with the additional cost consumption. For the actual welded structure with larger width, the transverse distance d1 will also be sensitive, which could not be far away from the welding line. Therefore, the transverse distance d1 of transient thermal tension is essential to effectively reduce the out-of-plane welding distortion of butt-welded joint, and there must be an appropriate range of transverse distance d1 to mitigate the welding buckling well.

Moreover, longitudinal plastic strains on middle cross-section of butt-welded joints with different transverse distances (d1) of induction heating

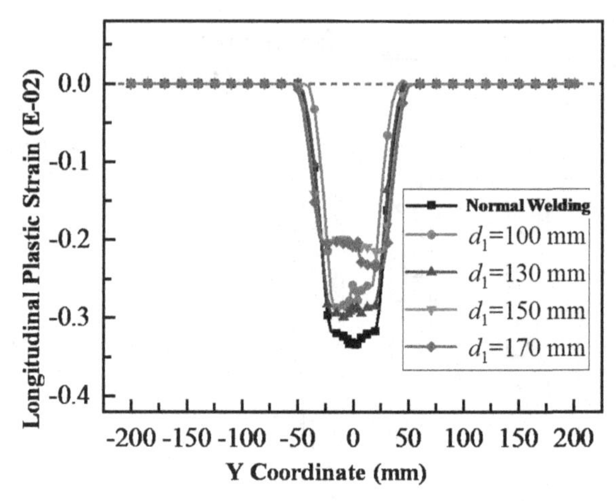

FIGURE 5.10 Comparison of longitudinal plastic strain on middle cross-section of butt-welded joints with different transverse distance (d1) of induction heating

were compared as demonstrated in Figure 5.10. There will be no additional longitudinal plastic strain in the heated region with induction heating because maximal temperatures (258°C) due to induction heating are less than the yield temperature. It can be seen that maximal compressive plastic strain in the region of welding line is significantly reduced with transient thermal tension compared to that with conventional welding. With the increasing transverse distance d1, the reduction effect of maximal compressive plastic strain will be obvious, which has identical behavior as demonstrated in Figure 5.10.

5.3 INFLUENCE OF OTHER PROCESS PARAMETERS

The influence of magnitude of induction heating and transverse distance d1 has the dominant role in reducing the out-of-plane welding distortion as well as welding buckling, in particular longitudinal plastic strain and longitudinal shrinkage force. The gap between induction heating coil and heated plate will also influence the temperature profile, which is fixed to be a constant of 8 mm based on induction heating experiment.

As shown in Figure 5.11, the influence of longitudinal distance d2 between induction heating coil and welding arc on reduction of out-of-plane welding distortion was examined by FE computation. In detail, the maximal temperature of heated region with induction heating is 258°C and the transverse distance d1 between induction heating coil and welding arc is 150 mm. There are eight cases with different longitudinal distance d2

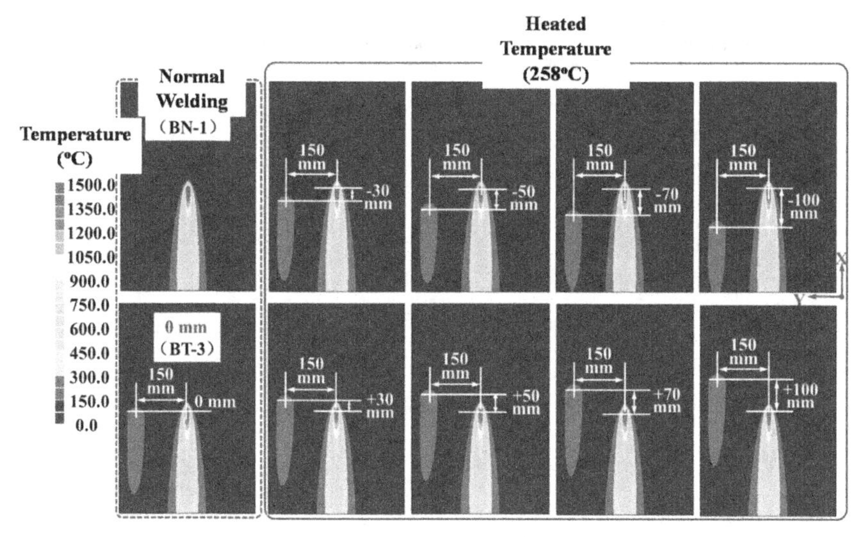

FIGURE 5.11 Computed temperature of butt welding with different longitudinal distance (d2) of induction heating

with ±30 mm, ±50 mm, ±70 mm, and ±100 mm, in which a positive value means induction heating will be applied before welding.

The transient temperature distributions with different longitudinal distance d2 were computed and compared as shown in Figure 5.12. Moreover, the computed out-of-plane welding distortions with different longitudinal distance d2 were then obtained as demonstrated in Figure 5.12. It can be seen that the longitudinal distance d2 also has an obvious influence on out-of-plane welding distortion. As shown in Figure 5.13, the maximal relative out-of-plane welding distortion and overall standard deviation for the butt welding with different longitudinal distance d2 were compared. In conclusion, maximal relative out-of-plane welding distortion and overall standard deviation for the butt welding are only 5.38 mm and 0.89 mm, which are reduced by 76.5% and 80.8%, respectively, compared to that with conventional welding, when the longitudinal distance d2 is 50 mm (induction heating is applied before welding arc with 50 mm distance).

Moreover, the influence of size of induction heating coil as well as the number of applied induction heating coils on out-of-plane welding distortion of examined butt-welded joint was also examined by means of thermal elastic plastic FE computation.

With the conditions of maximal temperatures of heated region with induction heating (258°C), transverse distance d1 (150 mm), and longitudinal distance d2 (0 mm), there are three cases with different sizes of

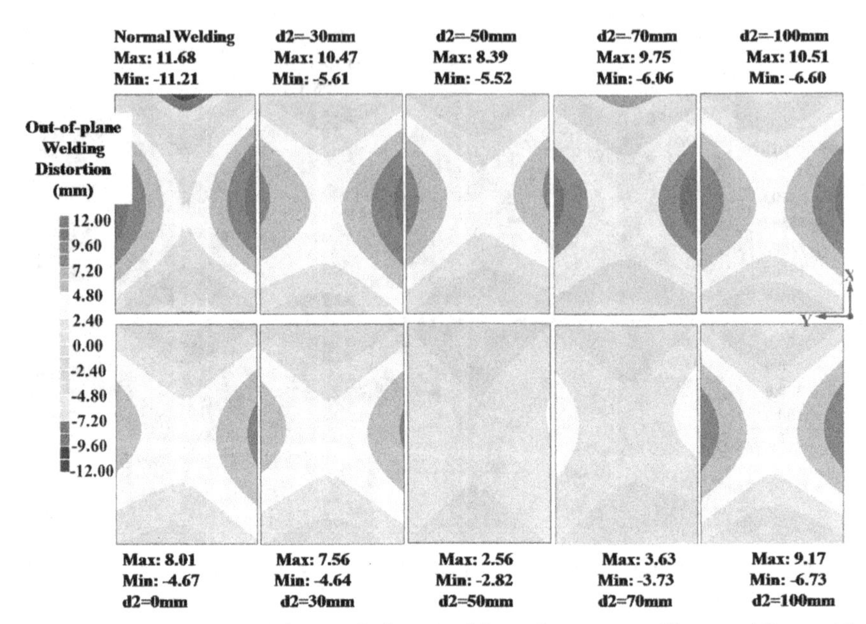

FIGURE 5.12 Computed out-of-plane welding distortion of butt welding with different longitudinal distance (d2) of induction heating

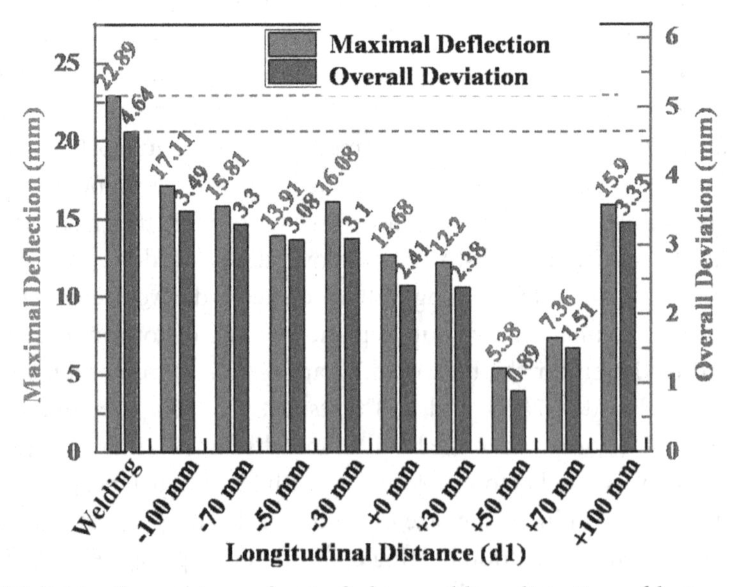

FIGURE 5.13 Comparison of out-of-plane welding distortion of butt welding with different longitudinal distance (d2) of induction heating

induction heating coil that were numerically examined. The size of induction heating coil for butt-welded joint of BT-3 is 72 mm. As shown in Figure 5.14, temperature contours as well as maximal temperature distribution of thermal analysis with different sizes of induction heating coil were demonstrated. It also can be seen that the induction heating has less influence on the temperature profile in the region of welding line.

As shown in Figure 5.15 of comparison of out-of-plane welding distortion with different sizes of induction heating, the out-of-plane welding distortion will reduce with the increasing size of induction heating. In

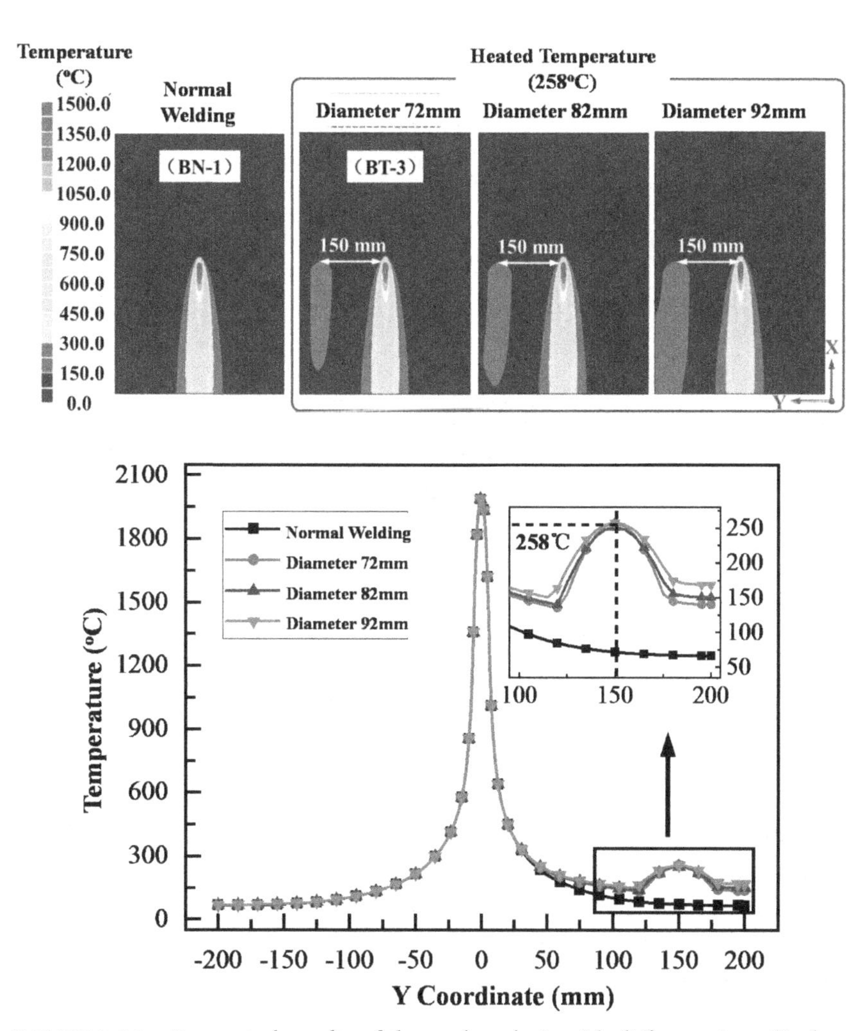

FIGURE 5.14 Computed results of thermal analysis with different size of induction heating coil

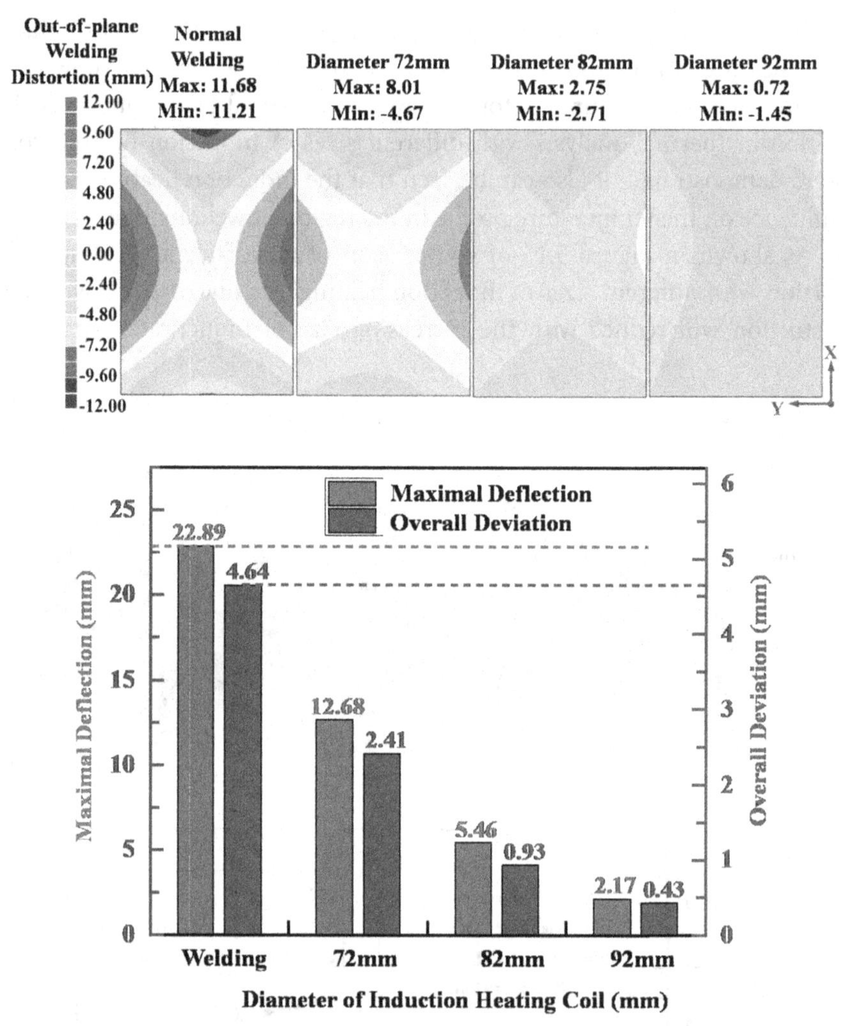

FIGURE 5.15 Computed out-of-plane welding distortion with different size of induction heating coil

addition, when the size of induction heating coil is 92 mm, maximal relative out-of-plane welding distortion was reduced to 2.17 mm by 90.5% compared to 22.89 mm with conventional welding, and overall standard deviation was reduced by 90.7% from 4.64 mm with conventional welding to 0.43 mm with thermal tension technique. It can be seen that the out-of-plane welding distortion was almost reduced and perfect butt-welded joint with accurate fabrication was obtained.

Furthermore, the influence of two induction heating coils on out-of-plane welding distortion was eventually examined. In detail, the transverse distance d1 and longitudinal distance d2 of two induction heating coils were 150 mm and 0 mm, respectively. The maximal temperature of heated region with induction heating is 258°C. With thermal analysis, the transient temperature contours as well as maximal temperature distribution

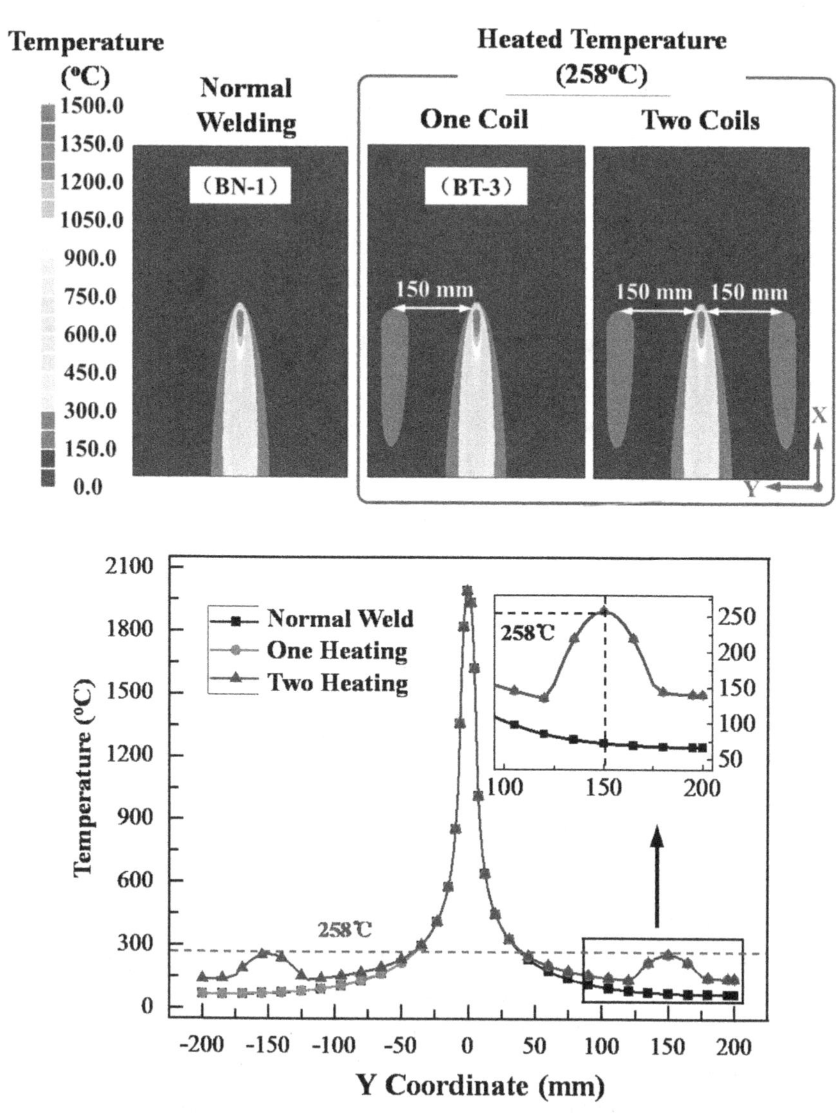

FIGURE 5.16 Computed results of thermal analysis with different numbers of applied induction heating coil

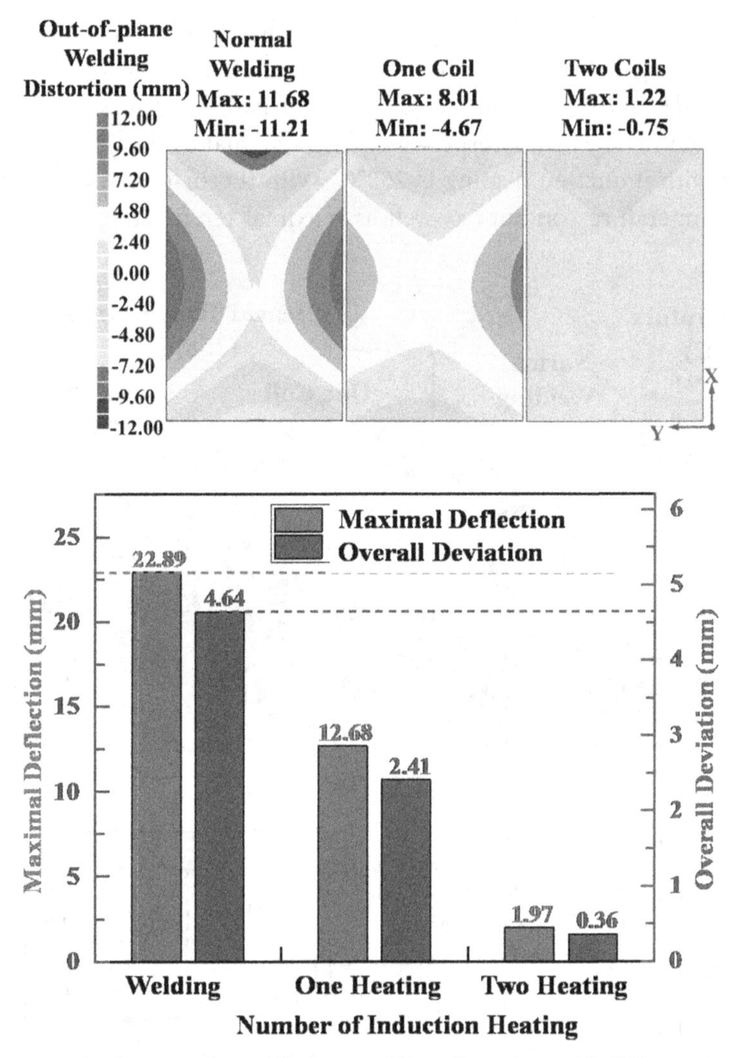

FIGURE 5.17 Computed out-of-plane welding distortion with different numbers of applied induction heating coil

with different numbers of induction heating coils were computed as shown in Figure 5.16, while the induction heating has less influence on the temperature profile in the region of welding line.

As shown in Figure 5.17, comparison of out-of-plane welding distortion with different numbers of applied induction heating coils was demonstrated. In general, thermal tension with two induction heating coils has better reduction effect on out-of-plane welding distortion than that with one induction heating coil. When the two induction heating coils are applied, maximal relative out-of-plane welding distortion and overall standard deviation for the butt welding are only 1.97 mm and 0.36 mm, which are reduced by 91.4% and 92.2%, respectively, compared to that with conventional welding.

5.4 SUMMARY

Parameters of thermal tension with induction heating have a significant influence on mitigation of welding buckling during the fabrication of thin plates. The influence of heating intensity as well as maximal heated temperature, transverse distance d1 between welding torch and induction heating coil, longitudinal distance d2 between welding torch and induction heating coil, diameter of induction heating coil, and number of induction heating coil were all examined by FE computation.

CHAPTER 6

Fillet Welding Buckling Experiment

AFTER INVESTIGATION OF BUTT-WELDED joint of AH36 thin plate by means of experiment and FE computation, another typical welded joint with fillet welding was then experimentally examined, while welding without and with thermal tension caused by induction heating were both considered to prevent twisting buckling distortion.

6.1 FILLET WELDING PROCEDURE AND BEHAVIOR

The material of examined fillet welded joint is ship steel AH36, and the geometrical size and welding groove were both indicated in Figure 6.1. In detail, the flange plate was designed in longitudinal direction with 600 mm and transverse direction with 400 mm, the web plate was considered in longitudinal direction with 600 mm and height direction with 200 mm, and all plates are 5 mm in the thickness direction. Welded groove with single V shape for web plate was employed and the groove angle is about 45°, while there is no gap between flange and web plates for full penetration welding.

Similar to above butt welding, MAG (metal active gas) Arc welding method was employed, while the shielding gas is the mixed gas with 80% Ar and 20% CO_2 and NB-350 welding equipment and SB-10-350 wire feeder were used again. Moreover, solid wire of ER50-6 with similar strength compared with ship steel of AH36 was employed, while the diameter of solid wire is 1.2 mm.

DOI: 10.1201/9781003442523-6

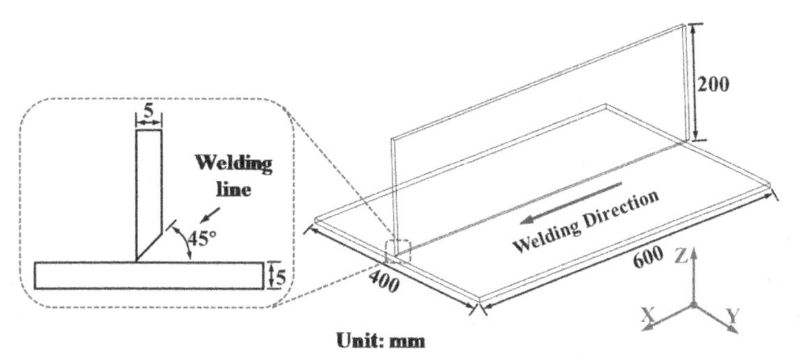

FIGURE 6.1 Geometrical size and welding groove of fillet-welded joint

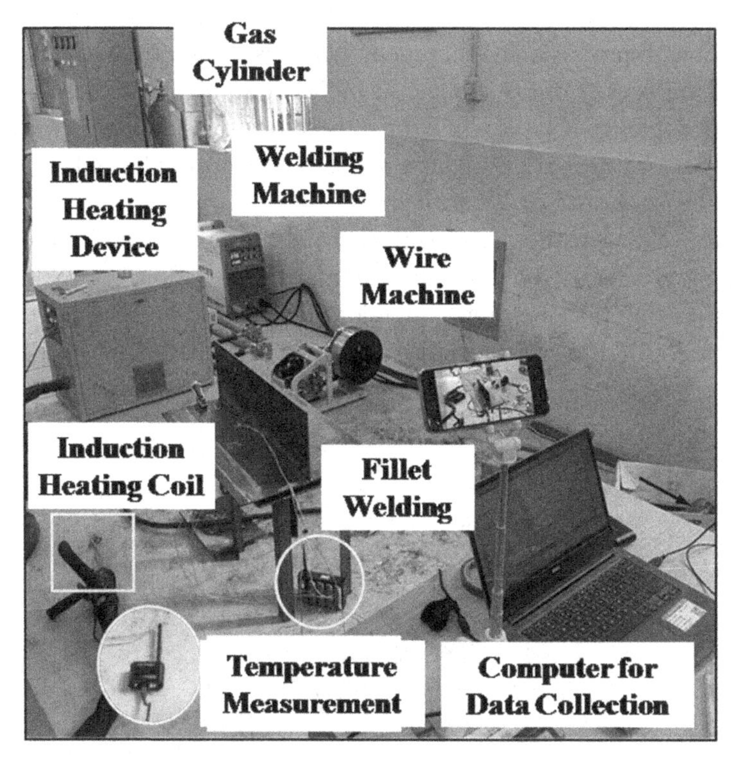

FIGURE 6.2 Experimental platform for fillet welding of AH36 thin plate

As shown in Figure 6.2, experimental platform for fillet welding of AH36 thin plate was employed again, which also includes welding equipment, induction heating equipment, temperature measurement, and data transfer and storage system. According to the above investigation of thermal tension caused by induction heating, the diameter of induction heating coil is about

52 mm, and high-temperature-resistant tape with material of refrasil glass fiber was still employed to fix a ceramics filler piece with thickness of 8 mm under the induction heating coil, which will make sure the constant gap between induction heating coil and steel plate, as well as identical output power of induction heating during the entire welding experiment.

Before the actual welding, grinding machine will be employed to polish the welded groove of web plate until the welded groove satisfies the requirement of design of welded groove. Then, the web plate will be placed and fixed to the appropriate position on flange plate with tack welding, while the tack welding was carried out at the back location of fillet welding line as indicated in Figure 6.3. After assembling of web and flange plates, K-type thermocouple of platinum-rhodium alloy was fixed on the top and back surfaces with professional equipment, which is similar to the case of butt-welded joints as examined above. As indicated in Figure 6.4, there are five thermocouples of K type, while three thermocouples of K type were fixed on the back surface of flange plate and two thermocouples of K type were fixed on the top surface of flange plate.

Fillet welding without and with thermal tension caused by induction heating was all carried out by means of manual operation. For the thermal tension during fillet welding, induction heating will be conducted at the opposite location of fillet welding and there is a web plate to support an

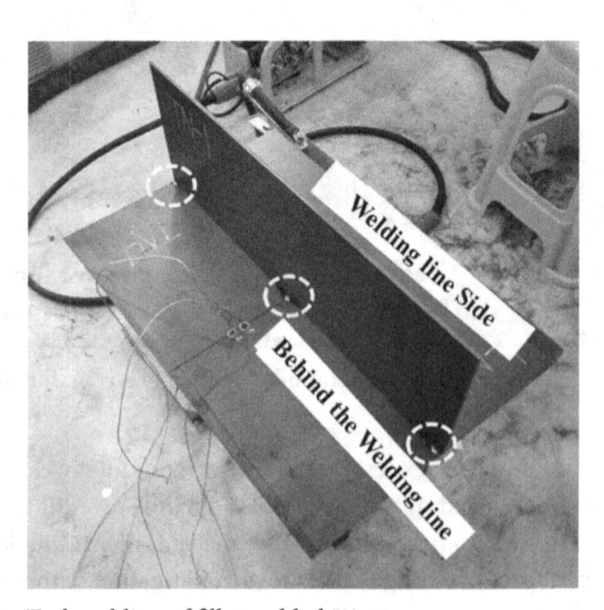

FIGURE 6.3 Tack welding of fillet-welded joint

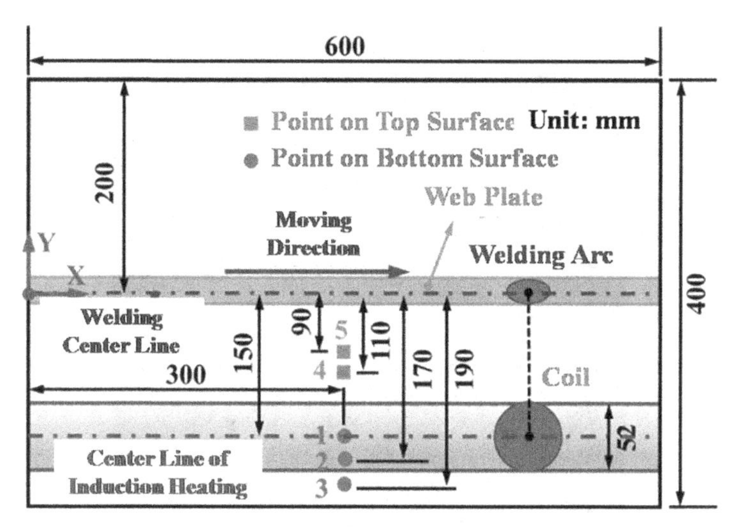

FIGURE 6.4 Design of temperature measurement and thermal tension during fillet welding

TABLE 6.1 Parameters of welding and induction heating for fillet welding.

Fillet Welding No	TN-1	TT-1	TT-2
Welding Procedure	Normal	With Thermal Tension Due to Induction Heating	
Current (A)	286~289	285~295	285~292
Voltage (V)	27.5~27.7	27.5~27.7	27.5~27.7
Welding Time (s)	130	135	133
Moving Speed (mm/s)	4.6	4.4	4.5
Current of Induction Heating (A)	/	600	400
Transverse Distance d1 (mm)	/	150	150
Longitudinal Distance d2 (mm)	/	0	0

effective shield against welding arc as shown in Figure 6.4. In detail, the transverse distance between welding line and center region of induction heating is 150 mm, and the longitudinal distance between them is 0 mm, which means induction heating coil will move with welding arc simultaneously. The difference for fillet welding with thermal tension is the reached maximal temperature due to induction heating, as well as applied output current of 600 A with TT-1 and 400 A with TT-2. In general, there will be higher maximal temperature caused by induction heating with larger output current of induction heating.

As summarized in Table 6.1, there are all welding conditions and parameters of induction heating for comparative fillet welding experiment,

FIGURE 6.5 Fillet welding process of AH36 without and with thermal tension

FIGURE 6.5 (Continued)

and there is an accepted tolerance of welding time as well as welding speed due to manual welding process. Thus, the fillet welding with thermal tension can be considered based on the normal welding with identical welding heat input.

As demonstrated in Figure 6.5, fillet-welded joints of AH36 thin plate were assembled without and with thermal tension. For the fillet welding without thermal tension, only one experienced welder achieved the whole welding process; meanwhile, during the fillet welding with thermal tension, one experienced welder conducts the welding while another experienced welder handles induction heating coil with simultaneous movement according to welding arc to heat appointed base material of flange plate. In addition, the moving speed of three fillet welding experiment will be consistent with the control of many marks on the welding line. After welding completion and cooling down to room temperature, the examined fillet welded joints were placed together as demonstrated in Figure 6.6.

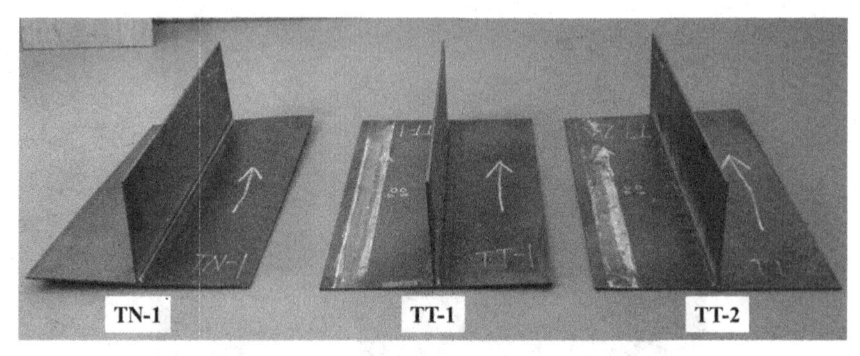

FIGURE 6.6 Experimental fillet-welded joints after cooling down

6.2 TEMPERATURE MEASUREMENT WITH THERMOCOUPLE

In order to examine the effect of induction heating during thermal tension on welding temperature field of fillet-welded joint with AH36 thin plate, there are several thermocouples fixed on surface of flange plate as mentioned in Figure 6.7 to obtain their transient temperatures during welding process. Similar measurement system of temperature during butt-welding experiment was again employed, which generally includes temperature record equipment of JM3818A as illustrated in Figure 6.2 and K-type thermocouple of platinum-rhodium alloy as displayed in Figure 6.7.

All measured points with K-type thermocouple were fixed with professional machine on the surface of flange plate opposite to the fillet welding. In addition, measured points of 1–3 are located at the back surface of flange plate just under the heated region of induction heating, and measured points of 4–5 are located at the top surface of flange plate between welding line and induction heating area.

As a result, measured maximal temperatures of designed poisons during fillet welding experiment without and with thermal tension caused by induction heating were compared as shown in Figure 6.8. The measured maximal temperatures during fillet welding with thermal tension in the cases of TT-1 and TT-2 are all higher than that during fillet welding without thermal tension in the case of TN-1. In particular, measured maximal temperatures of positions 1 and 2, which are located at the back surface of flange plate just under the region of induction heating, have obviously incensement compared with that in the case of TN-1. In detail, the maximal temperatures of position 1 are 357°C and 203°C for the case of TT-1 and TT-2 with thermal tension, and which is only 69°C for the case of TN-1.

FIGURE 6.7 Position of thermocouple on the fillet-welded joint

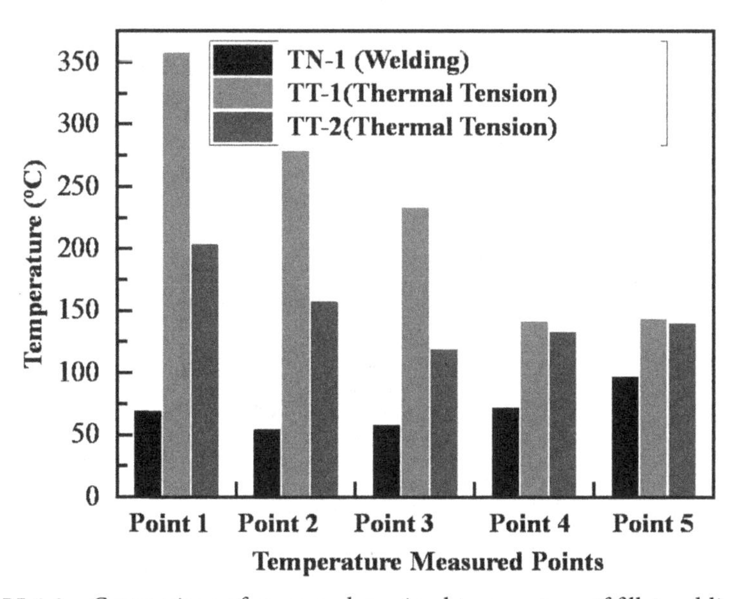

FIGURE 6.8 Comparison of measured maximal temperature of fillet welding

6.3 OUT-OF-PLANE WELDING DISTORTION MEASUREMENT

With the three-coordinate measured machine supported by Hexagon Company as demonstrated in Figure 3.20, out-of-plane welding distortions of examined fillet-welded joints after cooling down to room temperature were measured. Due to the operation limitation and existence of web plate, the measured positions as demonstrated in Figure 6.9 were considered while the region between web and flange plates still could not be measured.

In particular, out-of-plane welding distortion as well as twisting buckling of examined fillet welded joint is obvious and will be paid more attention. For both flange and web plates, the measurement distance with

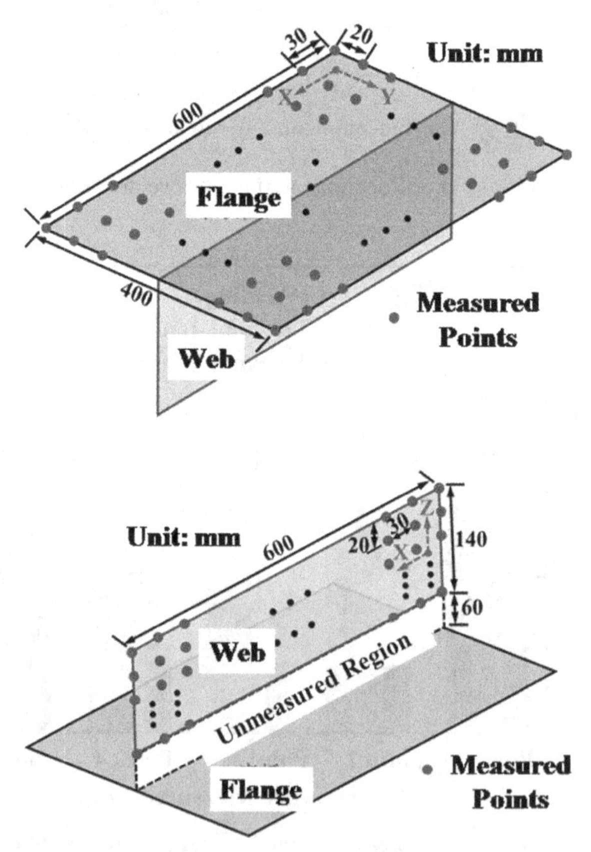

FIGURE 6.9 Out-of-plane welding distortion measured plan of fillet-welded joint

FIGURE 6.10 Measurement of out-of-plane welding distortion of fillet-welded joint with three-coordinate measured machine

probe of three-coordinate measuring machine is about 30 mm in longitudinal direction and 20 mm in transverse direction. In order to reduce the measurement tolerance, measured surfaces of examined fillet-welded joints should be cleaned to remove steel rust as well as welding spatter in advance. As demonstrated in Figure 6.10, examined fillet welded joints were placed on the platform with clamping to ensure the measured accuracy. And measured probe will be controlled by computer with the regular route designed beforehand to obtain the coordinate of measured position. Measured coordinates of point cloud will be synchronously displayed with the corresponding software in computer.

The measured data with point cloud can be analyzed to be plotting contour of out-of-plane welding distortion of examined fillet-welded joints as shown in Figure 6.11. It can be seen that large twisting distortion with buckling feature was observed for fillet-welded joint without thermal tension (TN-1), while the maximal value of out-of-plane welding distortion of flange plate is about 44.20 mm. With the thermal tension caused by induction heating, the deformed mode of fillet-welded joint with twisting buckling could not be modified; however, the value of out-of-plane welding distortion of flange plate was effectively removed to be 18.86 mm for the case of TT-1 and 23.29 mm for the case of TT-2. Moreover, based on the parameters of induction heating, it may be seen that out-of-plane welding distortion of flange plate can be better controlled with larger reached maximal temperature as well as output current of induction heating during the fillet welding experiment.

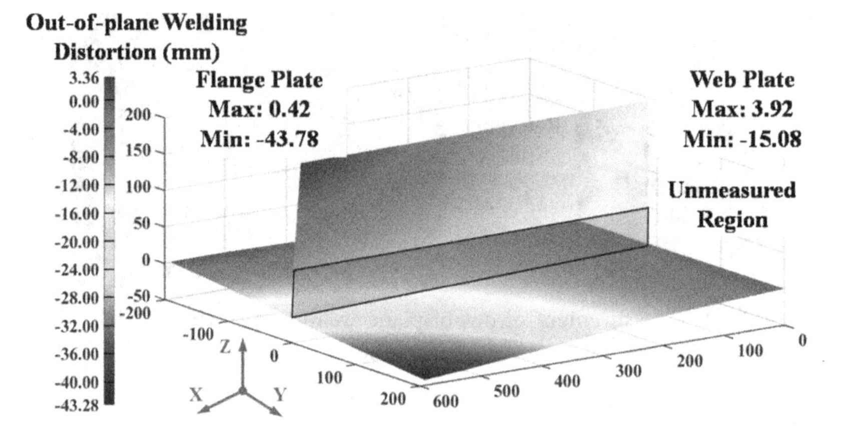

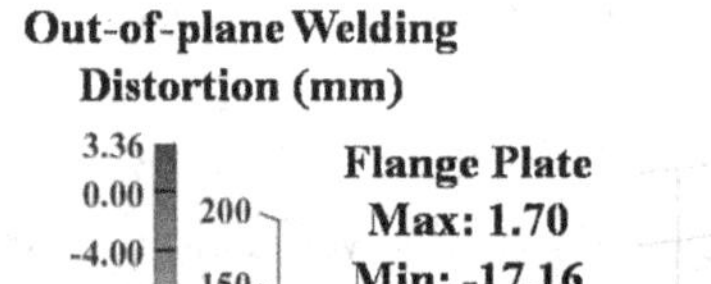

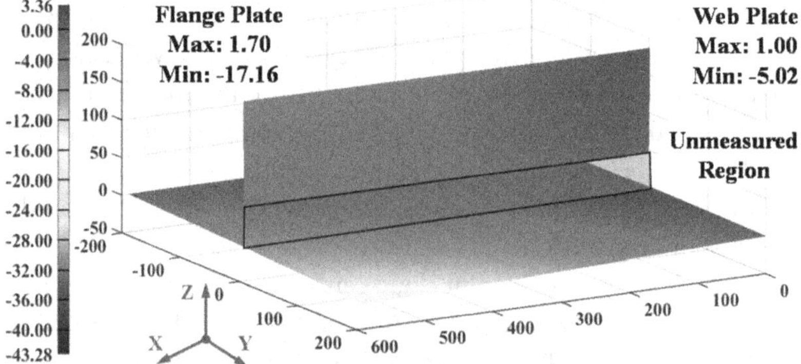

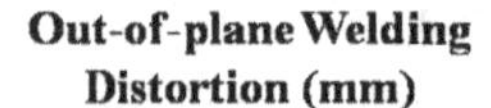

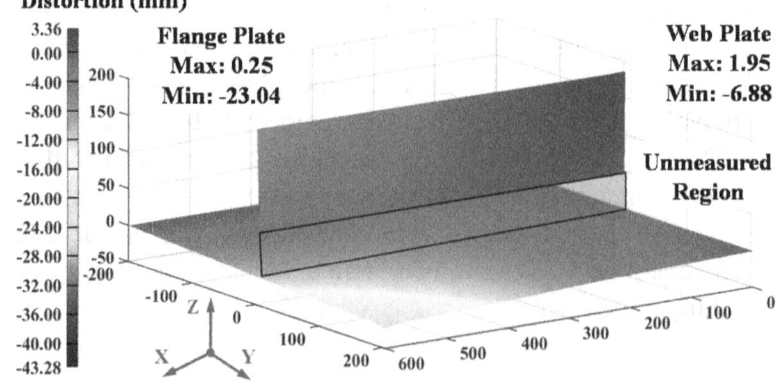

FIGURE 6.11 Measured result of out-of-plane welding distortion of examined fillet-welded joints

6.4 METALLOGRAPHY MEASUREMENT

After measurements of out-of-plane welding distortion, examined fillet-welded joints were cut by means of wire-electrode cutting with spark erosion to fabricate specimen of Metallography Measurement, while its position and dimension were both demonstrated in Figure 6.12. In detail, the specimen of Metallography Measurement is located in the middle region of welding line, while it has 70 mm in the width direction, 30 mm in the height direction, and 20 mm in the thickness direction.

In general, specimen of metallographic phase of examined fillet-welded joints should be processed by means of coarse grinding, precision grinding with abrasive paper for metallography, and burnishing, then reagent of iron trichloride was employed to corrode the cross-section of specimen of metallographic phase. After several minutes, specimen of metallographic phase will be washed with distilled water and wiped by means of absolute ethyl alcohol. With drying with air blower, the geometrical profiles of molten pool of examined fillet-welded joints can be eventually observed as shown in Figure 6.13. It can be seen that web plate was assembled with flange plate by full penetration welding and there is a certain depth of fusion due to molten pool in the flange plate. Moreover, there is neglectable difference in molten pool shape for examined fillet-welded joints, which can confirm that almost identical welding conditions were employed during fillet welding experiment.

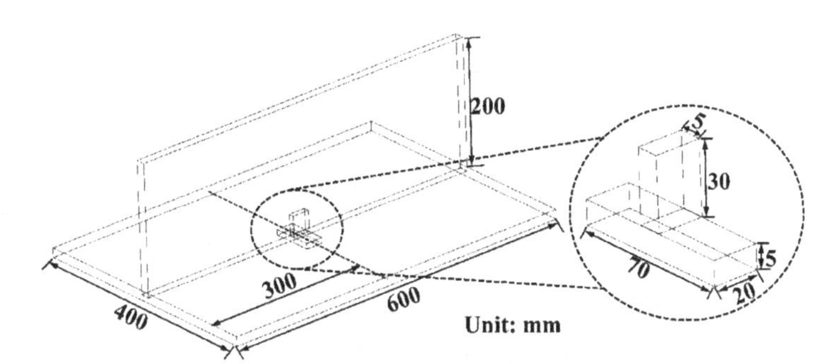

FIGURE 6.12 Position and dimension of specimen of Metallography of fillet-welded joint

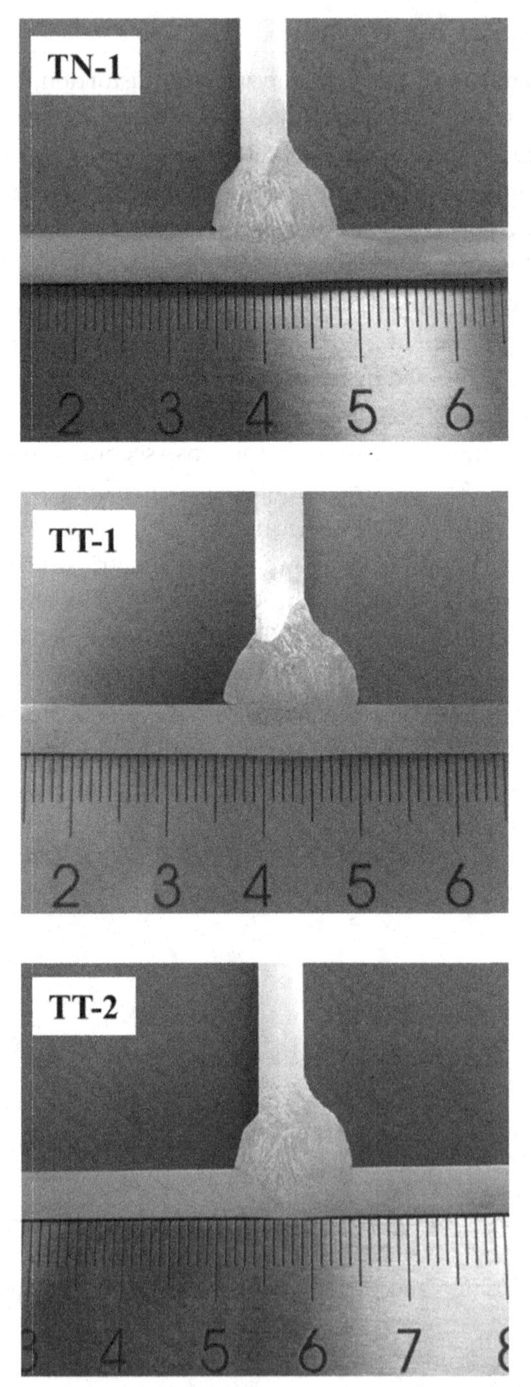

FIGURE 6.13 Metallography and molten pool shape of examined fillet-welded joints

6.5 SUMMARY

Fillet welding of AH36 thin plate with thickness of 5 mm was carried out with normal welding and thermal tension by means of induction heating. In addition, thermal cycles as well as maximal temperature of several points were measured by thermocouples, and out-of-plane welding distortion was measured by means of three-coordinate measured machine while twisting buckling was observed. Metallography and molten pool of examined fillet-welded joint was then measured and examined.

CHAPTER 7

Thick Plate Butt Welding and Measurement

HIGH TENSILE STRENGTH STEEL (HTTS) of Q690 is widely employed for offshore structures due to its perfect mechanical performance such as high ultimate strength and excellent fracture performance, which is usually produced by means of Thermomechanical Control Process (TMCP) with yield stress of 690 MPa. In addition, the microstructure of Q690 with TMCP is dominantly ferrite (F), while there is a little bit of existence of nonmetallic inclusion. As summarized in Tables 7.1 and 7.2, chemical components and mechanical properties of Q690 have been listed, respectively.

Compared with submerged arc welding (SAW) and shielded metal-arc welding with CO_2 shielding gas in the manufacturing of ship and offshore structures, manual electrode arc welding is usually employed for welding of high tensile strength steel such as Q690 and good-quality welded joint can always be obtained in actual fabrication. In order to match the mechanical performance of Q690, welding rod (E7618-G) of overseas product was considered, which is expensive, and its chemical component was summarized in Table 7.3. In addition, the diameter of considered welding rod is 4 mm.

DOI: 10.1201/9781003442523-7

TABLE 7.1 Chemical component of Q690.

C	Si	Mn	V	Ti	Cr	Ni	Nb	Cu	N	Mo	P+S+Al+N
0.165	0.11	1.8	0.05	0.04	1.0	0.8	0.021	0.8	0.015	0.3	0.085

TABLE 7.2 Mechanical properties of Q690.

Yield Stress	Ultimate Strength	Fracture Strength	Elongation	Strain Hardening Coefficiency	Strain Hardening Index
710 MPa	8630 MPa	572 MPa	0.23	1247 MPa	0.084

TABLE 7.3 Chemical component of welding rod (E7618-G).

C	Si	Mn	V	Ti	Cr	Ni	Nb	Cu	N	Mo	P+S+Al+N
0.06	0.8	1.5	0.1	-	0.3	3.5	0.04	0.2	-	0.75	0.06

7.1 BUTT WELDING PROCEDURE OF Q690 THICK PLATE

The examined butt-welded joint with HTSS Q690 is designed with thick plate, the thickness of which is 75 mm, and welding with full penetration was required for mechanical performance in service. Thus, welding groove to ensure the quality of butt-welded joint was considered with X shape due to its advantages such as high productivity, low cost of filler material, and control of out-of-plane welding distortion.

As shown in Figure 7.1, the detailed geometrical profile of butt-welded joint with Q690 thick plate was demonstrated, and the welding groove with an angle of 60°, gap of 2 mm, and root face of 3 mm was also indicated. With the wire-electrode cutting with spark erosion, welding groove was fabricated according to the design plan as shown in Figure 7.1. Before welding, steel rust and greasy dirt on the surface of welding groove should be removed, and two thick plates was assembled with tack welding as shown in Figure 7.2. In addition, preheating process with induction heating equipment on welding groove was also employed to prevent the occurrence of welding fracture as demonstrated in Figure 7.3, while the preheating temperature is about 150°C

As shown in Figure 7.4, welding scene of butt welding with Q690 thick plate was completely introduced. Moreover, welding rod (E7618-G) is always kept in the heat preservation barrels for perfect welding performance;

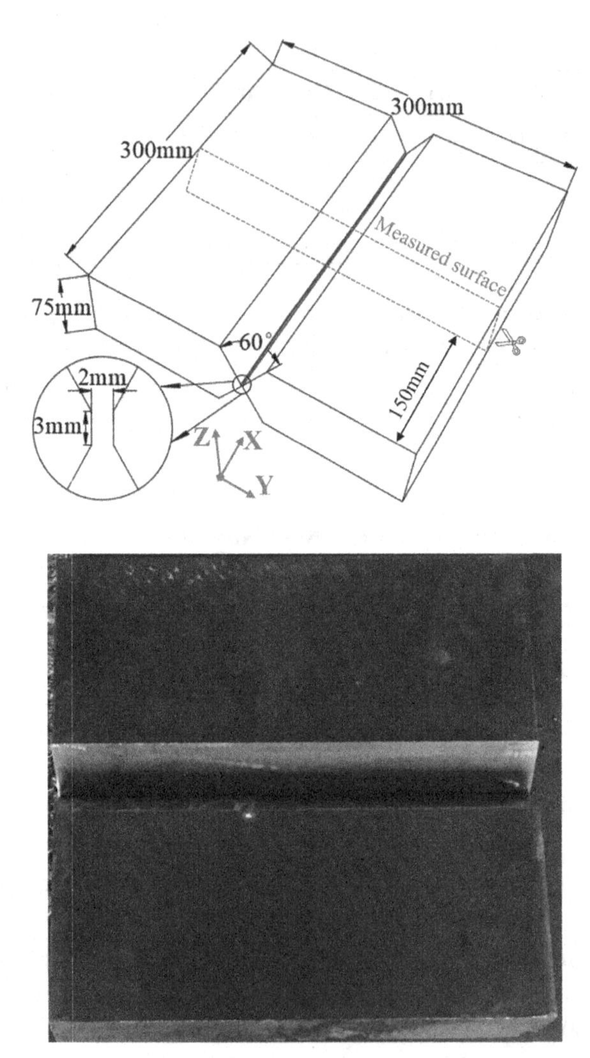

FIGURE 7.1 Dimension size and groove design of butt-welded joint with Q690 thick plate

manual arc welding machine with ZX7-400 IGBT of inverter direct current was employed as shown in Figure 7.5. After each welding, welding slag and spatter will be removed by means of mechanical practice such as iron brush and spade as demonstrated in Figure 7.5. And the interpass temperature during multipass welding of thick plate will be measured and controlled about 200~300°C

With the welded groove of X shape, out-of-plane welding distortion of butt-welded joint with Q690 thick plate can be effectively reduced by means

FIGURE 7.2 Tack welding of butt-welded joint with Q690 thick plate

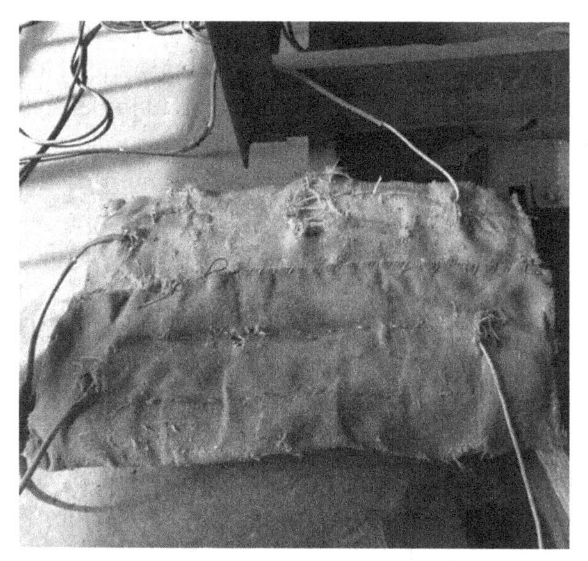

FIGURE 7.3 Preheating of welding groove with induction heating blanket

of symmetrical welding with lifting of rolling-over, while several displacement sensors were placed to monitor out-of-plane welding distortion.

During manual arc welding, molten drop filling with eight welding pass layers was finished from the top view beforehand. Then, butt-welded joint with Q690 thick plate was turned over by lifting machine and carbon arc

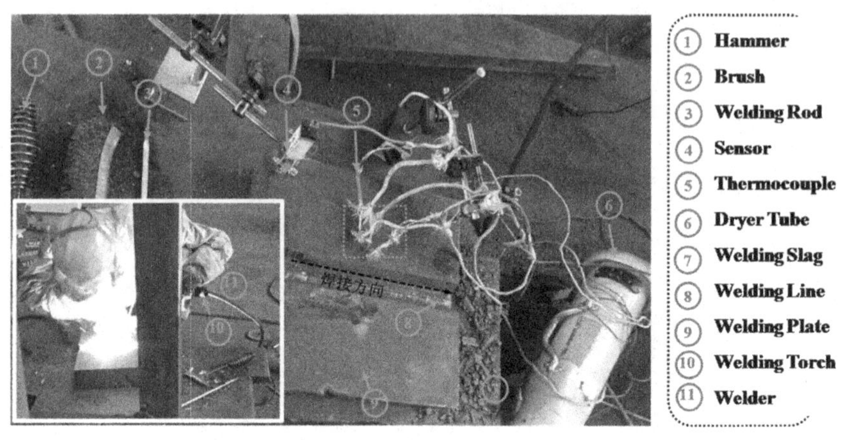

FIGURE 7.4 Overview of welding scene of butt welding with Q690 thick plate

gouging was employed to clean the already welded root region for later full penetration welding. After removement of iron brush and spade in the region of welding groove, molten drop filling with 12 welding pass layers was completed from the bottom view. Again, butt-welded joint with Q690 thick plate was turned over by lifting machine to initial status to achieve the remaining molten drop filling and cosmetic welding from the top view. Meanwhile, K-type thermocouple of platinum-rhodium alloy and wireless gathering system (JM3818A) was employed to measure the transient temperature as well as thermal cycle, while the detailed positions of thermocouples were designed as indicated in Figure 7.6. Thirdly, butt-welded joint with Q690 thick plate was turned over by lifting machine, and drop filling and cosmetic welding from the bottom view were achieved. As a result, there are 141 welding passes for examined butt-welded joint of Q690 thick plate in total, and each welding will consume about 100~150 s due to manual operation. The welding conditions were summarized in Table 7.4, and after cooling down to 200°C asbestos cloth was employed to enwrap the examined butt-welded joint of Q690 thick plate with 24 hours for hydrogen overflow treatment. After cooling down to room temperature, visual and ultrasonic inspections were carried out to check the welding defects such as slag inclusion, tiny flaws, lack of fusion, air holes, and undercut.

With the abovementioned butt welding of Q690 thick plate, it can be seen that manual welding of Q690 with thickness of 75 mm has some particular features such as complexity process, heavy workload, and difficult

FIGURE 7.5 Arc welding machine and clean of welding bead

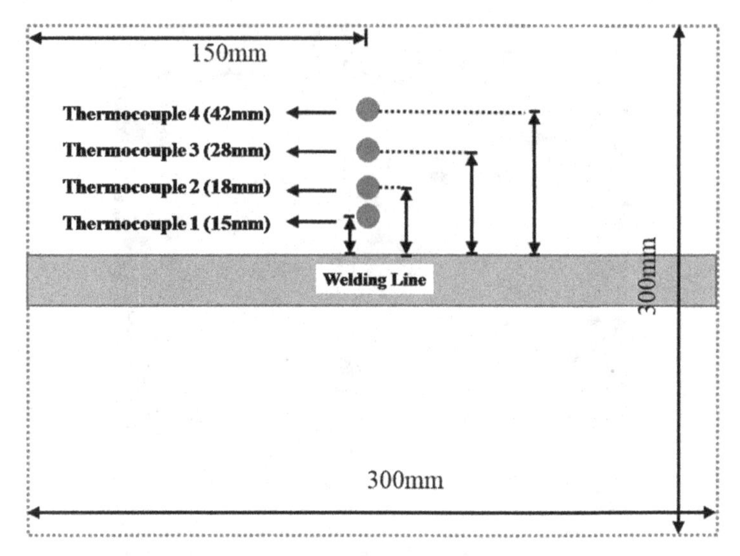

FIGURE 7.6 Position of K-type thermocouple

TABLE 7.4 Butt welding condition of Q690 thick plate.

	Current (A)	Voltage (V)	Welding Speed (mm/s)	Interpass Temperature (°C)	Heat Input (KJ/cm)
Root Welding	140~150	26~27	2.0~3.0	150	12.1~20.3
Filler Welding and Cap Welding	160~170	26~27	2.0~3.0	200~300	13.9~22.9

technique, which should be all solved to guarantee quality of welding process, mechanical performance of welded joint as well as fabrication cost.

Due to the multipass welding of butt-welded joint of Q690 thick plate, there are lots of thermal cycles. As shown in Figure 7.7, measured thermal cycles with thermocouples of No.1 as mentioned in Figure 7.6 were demonstrated for the left welding pass of last three welding layers from the top view. It can be seen that heat caused by welding arc will transfer to the location of thermocouple due to heat conduction, and the higher maximal temperature can be obtained when the welding pass is much closer to thermocouple. The measured maximal temperatures of thermocouples of No.1 are 261°C, 310°C, and 335°C, respectively.

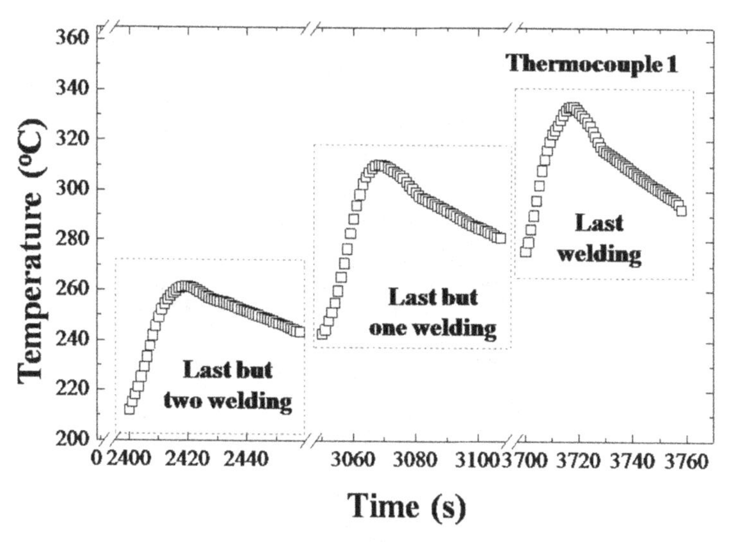

FIGURE 7.7 Measured thermal cycles during last three welding layers

7.2 RESIDUAL STRESS EVALUATION WITH CONTOUR METHOD

For the butt -welded joint of Q690 thick plate, longitudinal residual stress has complex distribution along the thickness direction, and tensile residual stress will result in stress concentration as well as crack propagation in the region with microdefect. Thus, accurate evaluation of welding residual stress in the butt-welded joint of Q690 thick plate is essential and desired. Generally, there are many practical approaches for welding residual stress measurement as mentioned above. And contour method is an ideal choice to obtain the longitudinal residual stress on the whole cross-section normal to welding line. Besides, due to the unstable condition of start and end of welding process, middle cross-section was considered for longitudinal residual stress measurement with contour method in the later investigation.

7.2.1 Cutting and Contour Measurement

Since the cutting quality of middle cross-section of butt-welded joint of Q690 thick plate has significant influence on the measurement accuracy of welding residual stress, machine of wire-electrode cutting with spark erosion (Sodick AQ400LS), as displayed in Figure 7.8, with lots of advantages of line cutting and high precision was employed to fabricate the cross-section for later contour measurement. In detail of wire-electrode cutting,

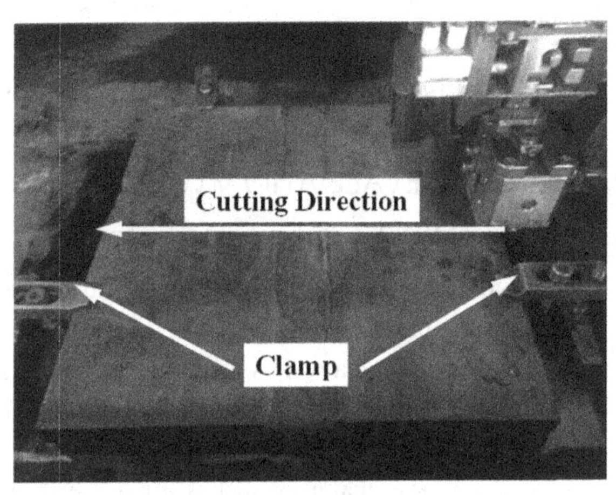

FIGURE 7.8 Wire-electrode cutting of examined butt-welded joint with Q690 thick plate

the cutting thread is copper wire with a diameter of 0.25 mm. Moreover, in order to prevent the influence of specimen movement on accuracy of contour measurement during wire-electrode cutting, butt-welded joint of Q690 thick plate was solidly fixed on the cutting platform as demonstrated in Figure 7.8.

During the wire-electrode cutting of butt-welded joint of Q690 thick plate, the whole butt-welded joint was submerged in the water, and the cutting speed was set as 0.5~0.8 mm/min to prevent stress generation due to cutting fabrication. After the wire-electrode cutting, cut surface of cross-section of butt-welded joint of Q690 thick plate should be cleaned and sealed to avoid the oxidation of metal surface, and there are two cut surfaces which will be marked with surfaces A and B for later individual measurement.

In the stage of contour measurement of cut surface, measured data is essential to evaluate the welding residual stress. Thus, release distortion after wire-electrode cutting is measured with three coordinate measuring machine with high precision supported by Hexagon as shown in Figure 7.9, while the measured precision is 1 μm and diameter of touching probe is 5 mm. During actual measurement, points with distance of 0.5 mm will be sequentially measured by touching method as demonstrated in Figure 7.9. Moreover, two cut surfaces with marks of A and B should be both measured to obtain the release distortion after wire-electrode cutting, while the measured sequence and direction should be consistent.

Due to the measured tolerance of release distortion of cut surfaces, it is necessary to carry out data analysis to remove the influence of tolerance disturbance caused by process of wire-electrode cutting and contour measurement.

7.2.2 Inverse FE Analysis for Residual Stress Evaluation

While data average and fitting were both considered as shown in Figure 7.10, original measured data of release distortion of cut surfaces A and B were individually displayed. It may be seen that there are some abnormal values of original measured data due to metal surface oxidation as well as measured tolerance. And these abnormal values will generate residual stress mutation in some local regions during evaluation of welding residual stress by means of inverse FE analysis.

Therefore, data analysis was carried out. Average release distortion of cut surfaces A and B was calculated to remove the influence of wire-electrode cutting beforehand. Then, fitting process with binary cubic spline surfaces was employed for smoothing of average release distortion, and eventual data as well as plotting contour was demonstrated in Figure 7.11. In addition, it can be seen that the plotting contour of release distortion after averaging and fitting is expressly smooth, and the root mean square (RMS) error value of fitting is only 0.001493 with acceptable reliability.

Finally, residual stress inversion analysis was carried out with FE computation. FE model of examined butt-welded joint with Q690 thick plate was made according to the cross-section of specimen, and there are 89,021 nodes and 87,010 elements in the FE model as shown in Figure 7.12.

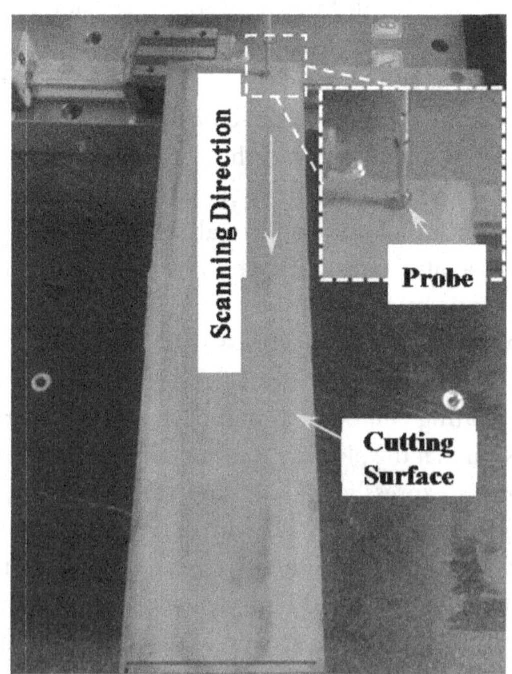

FIGURE 7.9 Contour measurement of cut surface of butt-welded joint with Q690 thick plate by three coordinate measuring machine

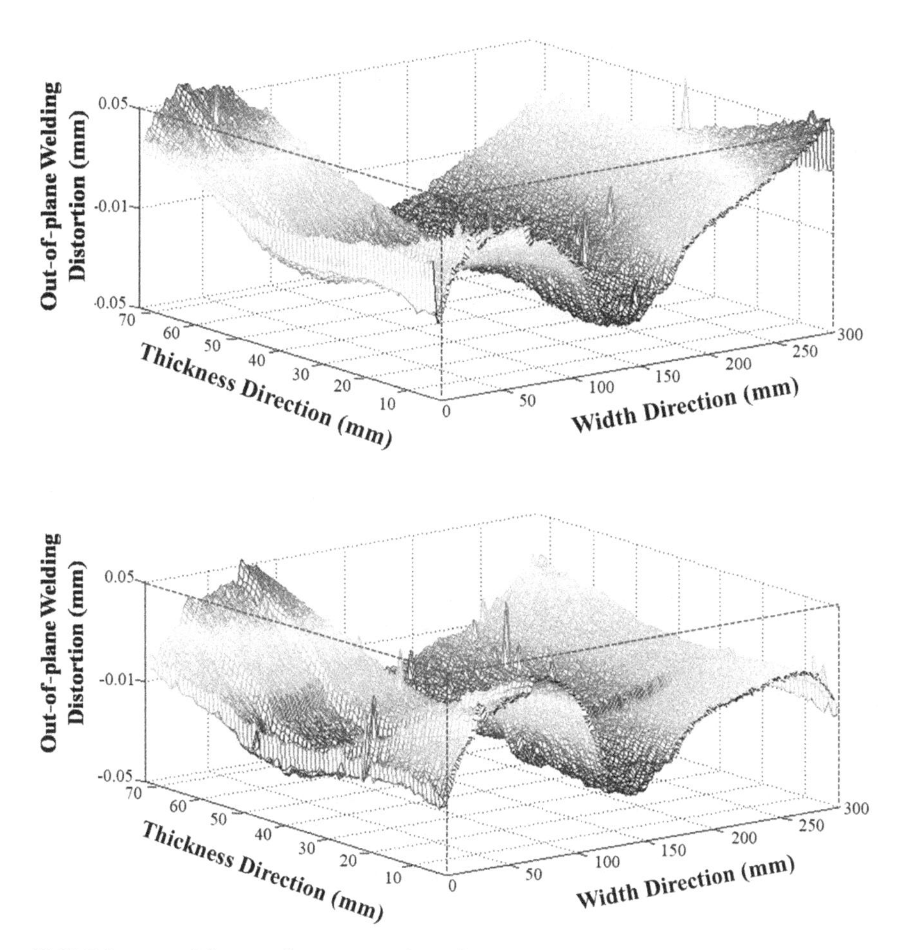

FIGURE 7.10 Measured contour value after wire-electrode cutting of butt-welded joint of Q690 thick plate

To prevent the rigid body motion, the boundary condition with black arrow during welding residual stress evaluation with measured release distortion was also indicated in Figure 7.12. Based on linear static mechanics analysis, longitudinal residual stress of middle cross-section of examined butt-welded joint with Q690 thick plate can be computed by means of FE computation by applying fitting data of contour measurement, while only elastic behavior of metal was considered with elastic modulus of 210 GPa and Poisson ratio of 0.3.

As shown in Figure 7.13, longitudinal residual stress of middle cross-section of examined butt welded joint with Q690 thick plate was plotted. There is discontinuous distribution of welding residual stress in the right

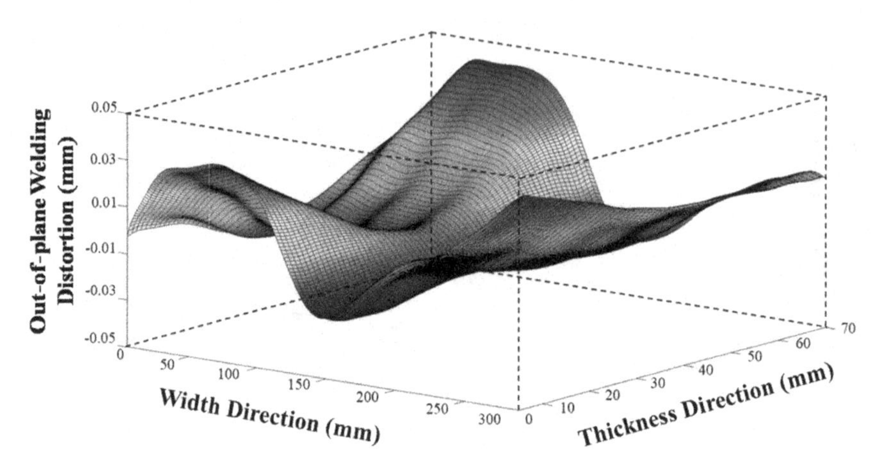

FIGURE 7.11 Contour value of middle cross-section after data analysis

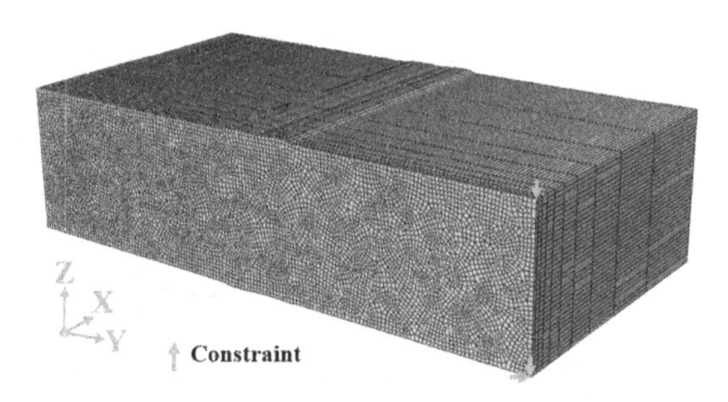

FIGURE 7.12 FE model of mesh for longitudinal residual stress evaluation by inversion analysis

region due to the transient break and replacement of copper wire during wire-electrode cutting. In addition, there is tensile residual stress with the maximal value of 200 MPa in the region of molten pool, which is much less than the yield stress of Q690. And the region located in the middle of molten pool in thickness direction has less tensile residual stress, which is marked with A as indicated in Figure 7.13. Abovementioned tensile stress in the region of molten pool is generated by constraint of surrounding base metal on welding line during cooling process. Compressive residual stress then occurred in the region far away welding line to balance the tensile residual stress. In the region of base metal, there will be compressive residual stress in the position near top and bottom surfaces and tensile residual stress in the middle position which is marked with B along the thickness direction.

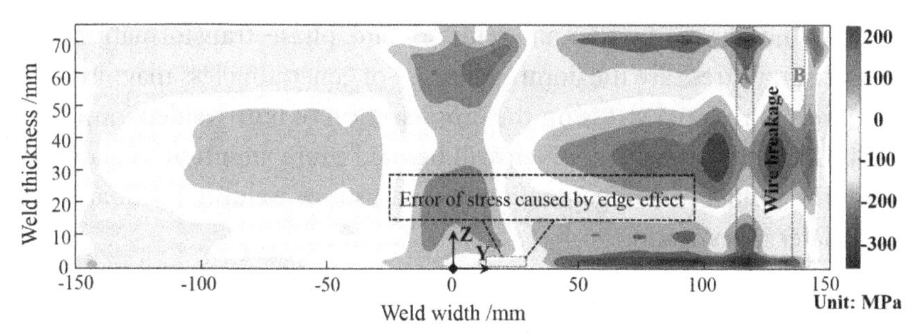

FIGURE 7.13 Longitudinal residual stress on middle cross-section of butt-welded joint of Q690 thick plate evaluated by contour method

The distribution and magnitude of evaluated longitudinal residual stress can be understood based on the actual welding process of butt-welded joint with Q690 thick plate, as well as wire-electrode cutting with spark erosion and contour measurement with three coordinate measuring machine. In particular, corresponding and possible influential factors were summarized and introduced as follows:

(1) There are solid phase transformations such as Martensite and Bainite in the region of molten pool and HAZ during cooling process, which will induce volume expansion to reduce the magnitude of tensile residual stress.

(2) Q690 thick plate with thickness of 75 mm is usually produced by Thermomechanical Control Process (TMCP), which will generate temperature gradient as well as the rolling residual stress in the thickness direction. This kind of rolling residual stress is considered initial stress before welding and results in the generation of tensile residual stress in the middle position of base metal in the thickness direction.

(3) Mechanical vibration due to welding scarfing cinder, welded joint transport will have little influence on welding residual stress relaxation.

(4) Removement of welded root region by means of carbon arc gouging has little influence due to heat treatment effect on welding residual stress relaxation.

(5) Inherent data tolerance is caused by wire-electrode cutting with spark erosion and contour measurement with three coordinate measuring machine.

With the above discussion, welding solid phase transformation and rolling initial stress are the dominant cause of generating less magnitude of longitudinal residual stress on the cross-section of butt-welded joint with Q690 thick plate together, which will be paid more attention to clarify in detail the influence and generation mechanism of welding residual stress in the later investigation.

7.3 METALLOGRAPHY AND HARDNESS MEASUREMENT

7.3.1 Metallography of Thick Plate Welded Joint

After cooling down to room temperature, wire-electrode cutting with spark erosion was employed again to make the metallography specimen of butt-welded joint of Q690 thick plate, while its position and dimension are demonstrated in Figure 7.14. In detail, the metallography specimen is located in the middle region of welding line, and its dimension is 20 mm in length direction, 60 mm in width direction, and 75 mm in thickness direction. By grinding with abrasive paper for metallography and corrosion with reagent of iron trichloride, metallographic phase of specimen of butt-welded joint of Q690 thick plate can be obtained by using a high-definition digital camera as shown in Figure 7.14.

In order to examine the microstructure with electron microscope, specimen of butt-welded joint of Q690 thick plate was polished again and corroded with 4% nitric acid ethyl alcohol, as shown in Figure 7.15. Ultraview deep microscope supported by Olympus was used to analyze the microstructure such as Martensite (M), Bainite (B), and Ferrite (F).

As shown in Figure 7.16 of metallography of butt-welded joint of Q690 thick plate, it can be clearly confirmed that the region of base metal, HAZ, and molten pool, and there are no welding defects such as slag inclusion, tiny flaw, lack of fusion, and air hole in the region of molten pool. In addition, the widths of molten pool are 42.4 mm, 5.9 mm, and 41.3 mm at the top surface, region of root welding, and bottom surface, respectively.

From Figure 7.17, the distributions of microstructure and its grain size in the HAZ are not uniform, while larger grain size is generated near the fusion line and smaller grain size appears far away from the fusion line. It can be understood that Austenite grain will rapidly grow due to higher maximal temperature of region near the fusion line. And there is almost no change of Austenite grain due to the lower maximal temperature of region far away from the fusion line. Besides, nonuniform distribution of

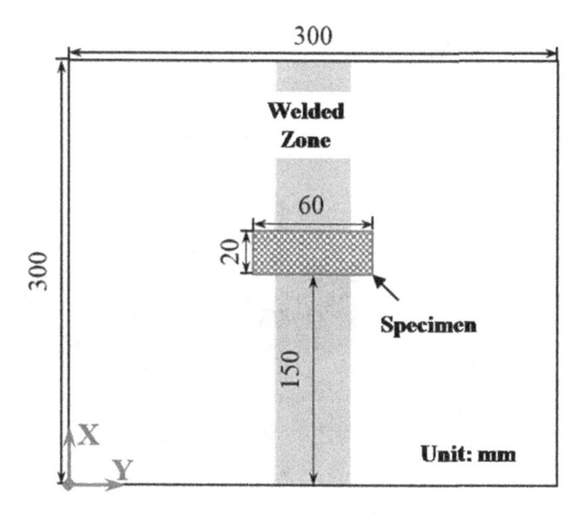

FIGURE 7.14 Metallography specimen of butt-welded joint of Q690 thick plate

grain size in HAZ is the basic cause of the weak mechanical performance of butt-welded joint. In general, ferrite will usually exist in the form of polygon grain, and Bainite will appear with graininess form in the region of molten pool. And Martensite will nucleate and grow to a lath-shaped form in the HAZ. Therefore, lath-shaped Martensite (M) is dominant microstructure of HAZ; graininess Bainite (B) and less Ferrite (F) are dominant microstructures of molten pool as demonstrated in Figure 7.17.

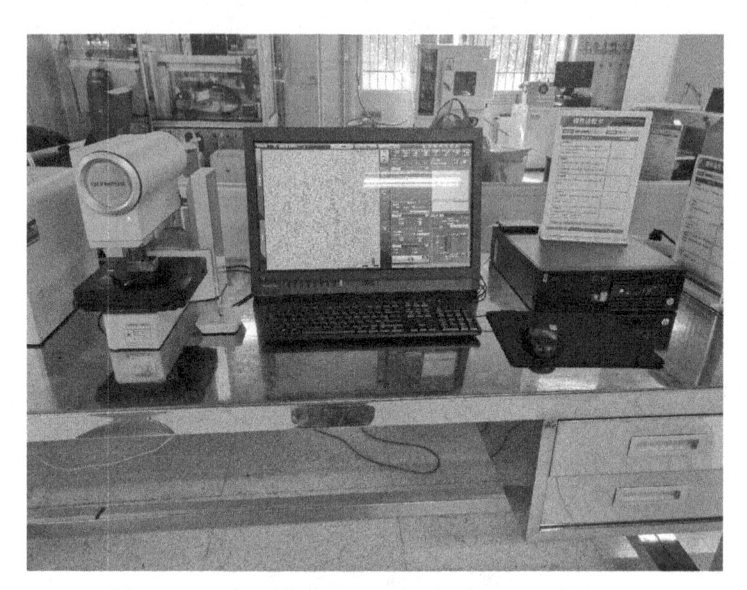

FIGURE 7.15 Microstructure observation with ultraview deep microscope

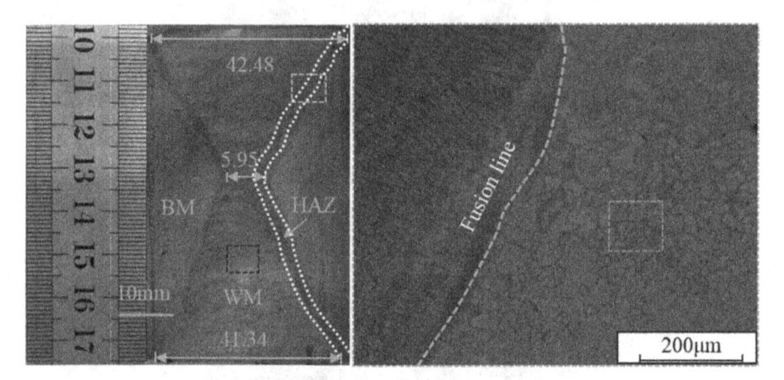

FIGURE 7.16 Geometrical profile of molten pool of butt-welded joint of Q690 thick plate (thickness: 75 mm)

7.3.2 Vickers Hardness Measurement

Hardness is a special performance and ability of material to counteract local deformation under external loading which is also obvious to present the hard grade of material. In addition, the hardness of material is dominantly determined by chemical component of metal, microstructure, and processing technique and therefore is essential index to analyze mechanical properties of toughness, abrasion, and strength of metals. In general, measurement of metal hardness can be achieved by the plunging method,

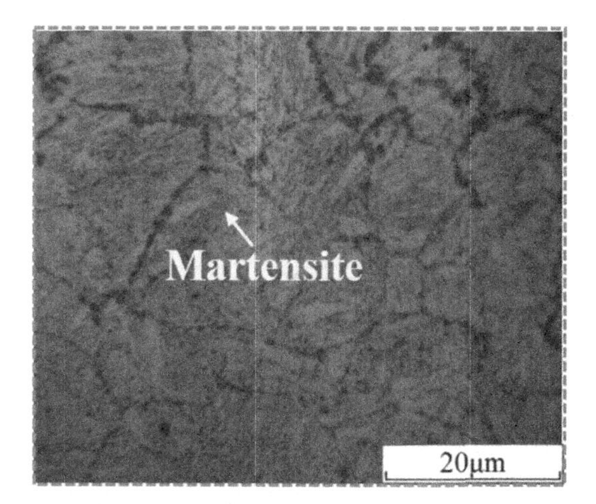

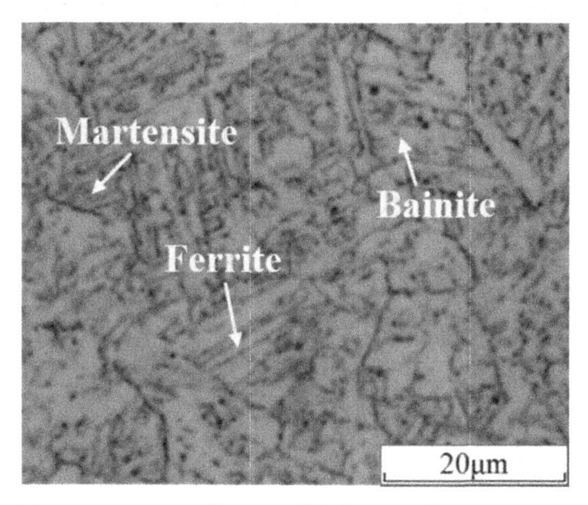

FIGURE 7.17 Microstructure of butt-welded joint of Q690

and according to the difference of squeeze head and pressure. There are three kinds of hardness: Brinell hardness (Unit: HB), Rockwell hardness (Unit: HR), and Vickers hardness (Unit: HV). Compared with limitation of Brinell hardness and Rockwell hardness for hard metal, Vickers hardness can be almost employed to present the hardness of all metals and measured results are regardless of applying pressure.

As shown in Figure 7.18, Vickers hardness can be measured and calculated by means of pressure per unit surface area of impression. Moreover, the conical angle of squeeze head is 136°, and squeeze head is plunged into the metal specimen with designed pressure during hardness test. Then external pressure can be unloaded after holding on for a certain time according to

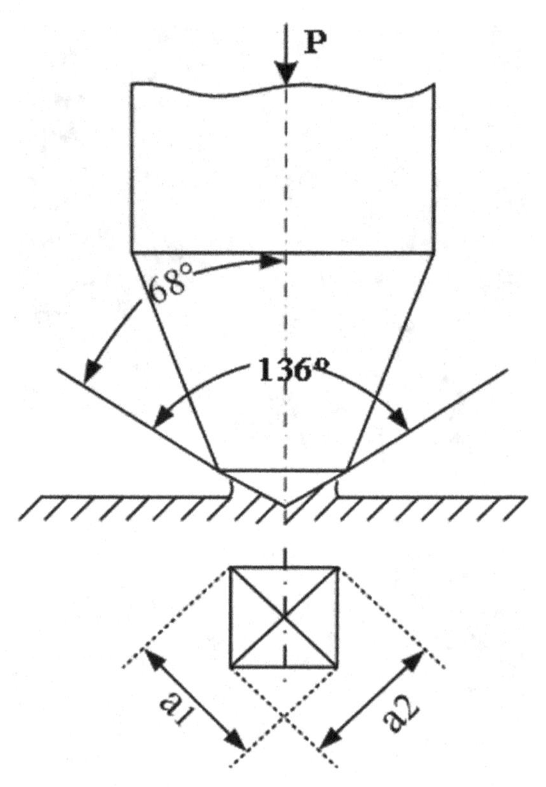

FIGURE 7.18 Image of Vickers Hardness Measurement

test specifications. After that, length of a1 and a2 as indicated in Figure 7.18 should be measured several times, and value of Vickers hardness can be calculated with Eq. (7.1).

$$HV = 0.189 \times \frac{P}{a_1 \times a_2} \tag{7.1}$$

where P means the external pressure on squeeze head with unit of N and a1 and a2 mean the length of diagonal lines of impression due to squeeze head with unit of mm.

As shown in Figure 7.19, Vickers hardness tester was employed to measure and evaluate hardness performance of examined butt-welded joint of Q690 thick plate. Moreover, the measured region of specimen was marked by means of dotted line with blue color for hardness measurement.

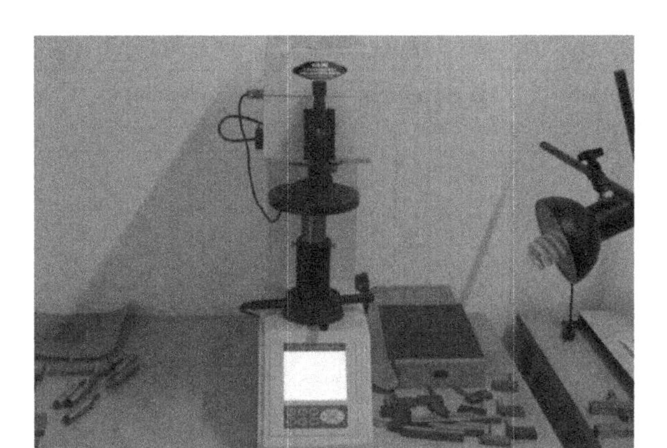

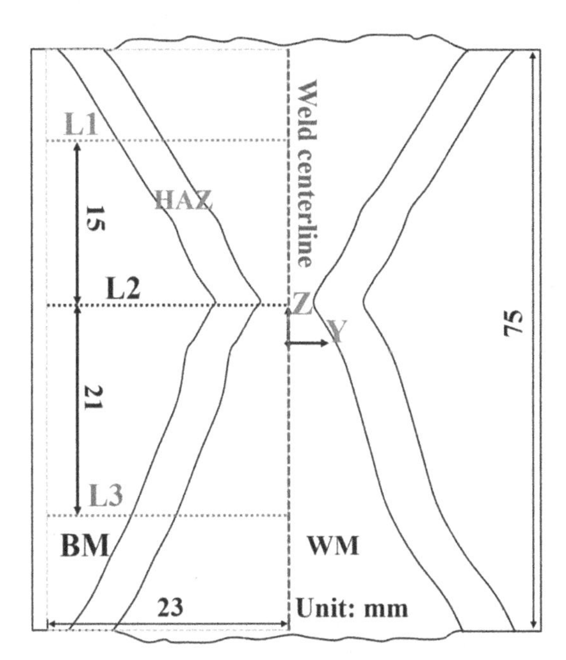

FIGURE 7.19 Measurement of Vickers hardness of examined butt-welded joint of Q690 thick plate

There are also three lines on the cross-section of the regions of top welding (line 1), root welding (line 2), and bottom welding (line 3) as demonstrated in Figure 7.19. In addition, the measured points on measured lines as mentioned above have a distance of 0.25 mm for the region of HAZ and distance of 1 mm for the region of base metal and molten pool.

During hardness measurement, the external pressure with 10 kg was applied on each measured position for 10 s, and measured Vickers hardness

TABLE 7.5 Measured hardness of butt-welded joint of Q690 (Unit: HV).

Distance to center of welding line (mm)	Measurement Line 1	Measurement Line 2	Measurement Line 3
−23	255.5	258.3	257.0
−22	251.7	244.9	260.8
−21	246.6	256.8	262.4
−20	254.0	258.3	255.5
−19	256.3	244.9	254.8
−18	239.1	256.8	262.4
−17	246.6	258.3	255.5
−16	254.0	244.9	254.1
−15	297.9	256.8	302.4
−14	329.8	255.1	335.8
−13	332.0	244.9	338.1
−12	277.9	258.3	288.4
−11	288.4	244.9	277.0
−10	288.4	256.8	283.1
−9	281.3	253.9	279.6
−8	279.6	256.9	287.5
−7	276.2	252.6	285.7
−6	294.8	256.9	294.8
−5	289.3	325.6	288.4
−4	289.8	346.6	283.1
−3	292.9	361.0	281.3
−2	289.6	285.5	283.1
−1	290.1	285.5	281.3
0	289.0	284.7	282.5

of examined butt-welded joint of Q690 thick plate has been summarized in Table 7.5. It can be seen that base metal has the lowest Vickers hardness with average value of 250 HV. Molten pool has the average value of Vickers hardness with 290 HV. The distribution of Vickers hardness can be understood with the microstructure in the region of base metal and molten pool. In detail, the microstructure of molten pool is dominantly Bainite (B). And for HAZ, the Martensite (M) is the dominant microstructure. In actuality, the Vickers hardness of microstructure will decrease with the sequence of Martensite, Bainite, ferrite, and austenite. Thus, HAZ has the larger Vickers hardness for the overall examined butt-welded joint of Q690 thick plate. Moreover, due to the bulky lath-shaped Martensite in the region near fusion line, the largest value of Vickers hardness can be observed with 361 HV.

7.4 SUMMARY

Butt-welded joint with Q690 thick plate was examined by welding experiment with multipass, while the thermal cycles were measured by means of thermocouple. Welding residual stress after cooling down was then measured by contour method by means of wire-electrode cutting with spark erosion and inverse FE analysis. Metallography and hardness of examined butt-welded joint were also measured and presented.

CHAPTER 8

Residual Stress Evaluations with FE Analysis

WITH THE ABOVE INVESTIGATION of contour measurement and inverse analysis, longitudinal residual stress on middle cross-section normal to welding line can be evaluated. In order to examine the welding residual stress of butt-welded joint with Q690 thick plate, transient thermal analysis considering heat conduction was carried out to obtain temperature profile as well as thermal cycles. Then, microstructure evolution of region of welded joint with temperature variation was investigated to consider solid phase transformation, which was confirmed with the comparison of hardness measurement. Eventually, welding residual stress was evaluated by considering the microstructure due to solid phase transformation, and initial stress condition due to TMCP (thermomechanical control process). Computed longitudinal residual stress was compared with the measured data by means of abovementioned contour method, and computed transverse residual stress will be employed to investigate the fracture performance of butt-welded joint with Q690 thick plate.

8.1 FE MODEL AND MATERIAL PROPERTIES

FE model with solid elements of butt-welded joint with Q690 thick plate has an essential influence on the temperature, microstructure transformation,

DOI: 10.1201/9781003442523-8

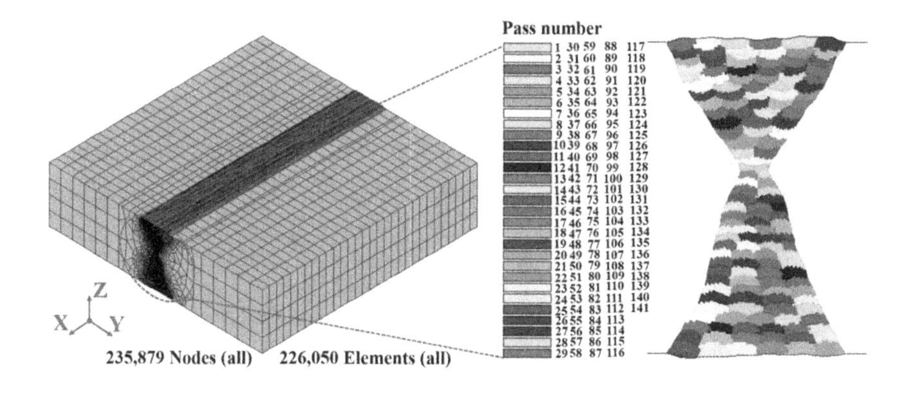

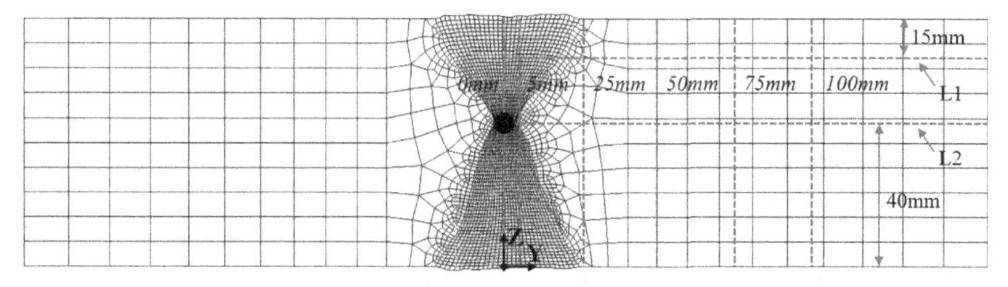

FIGURE 8.1 FE Model of butt-welded joint with Q690 thick plate

and welding residual stress. Due to the strong nonlinear feature of welded zone, fine mesh was usually employed, and coarse mesh was used for the region far away from the welding line. Thus, FE model of brick solid element with eight nodes was created in advance as shown in Figure 8.1, and there are 90,210 nodes and 85,290 elements of examined butt-welded joint with Q690 thick plate. In detail, the mesh size is only 1 mm in the welded zone and large element with size of 7.5 mm was considered in the region far away from the welding line. According to the actual welding experiment mentioned above, there are 141 welding passes as demonstrated in Figure 8.1 with different colors. With Figure 8.2 based on the chemical components, temperature-dependent material properties of Q690 as well as filler metal of (E7618-G) were employed in the later thermal and mechanical FE computations.

8.2 THERMAL ANALYSIS AND VALIDATION

With application of advanced computation techniques such as parallel computation, iterative substructure method (ISM), and element activity, an in-house code for thermal analysis of FE computation was employed

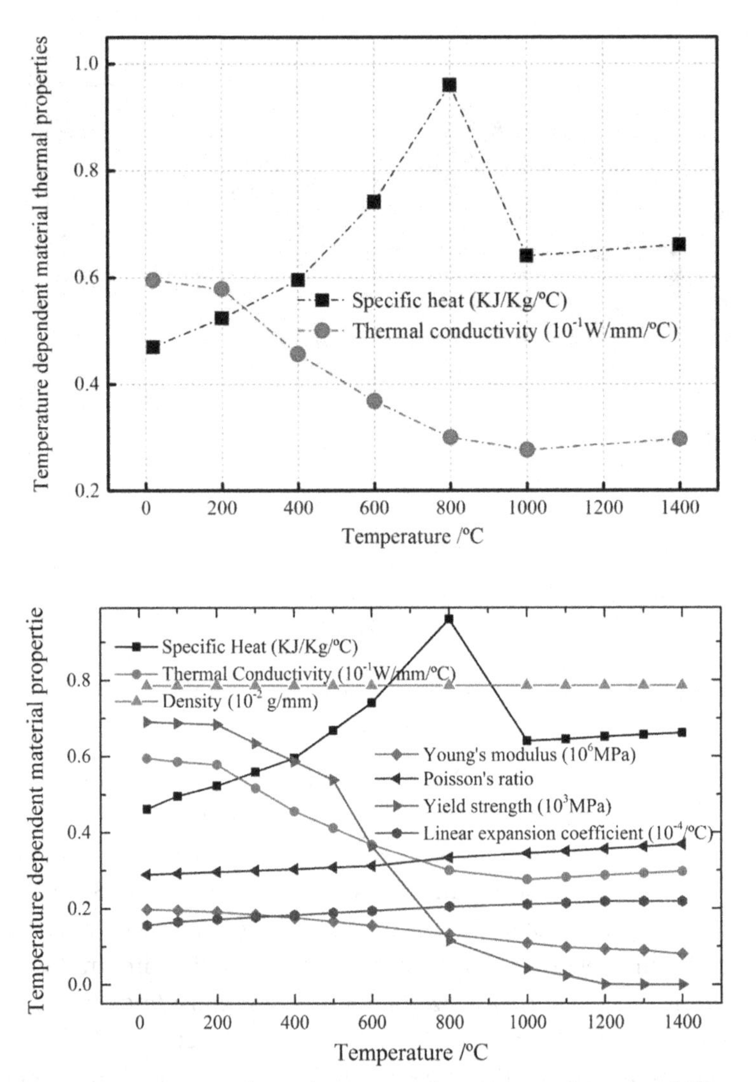

FIGURE 8.2 Temperature-dependent material properties of Q690 and filler metal

to simulate the transient temperature field of butt-welded joint with Q690 thick plate.

In addition, body heat source with uniform heat flow density was considered to model welding arc, which is determined by welding conditions as summarized in Table 7.4 such as welding current, welding voltage, welding speed, and thermal efficiency with value of 0.7 for manual welding. The time increment is considered to be 0.5 seconds, and initial temperature is about 150°C due to the preheating operation before welding.

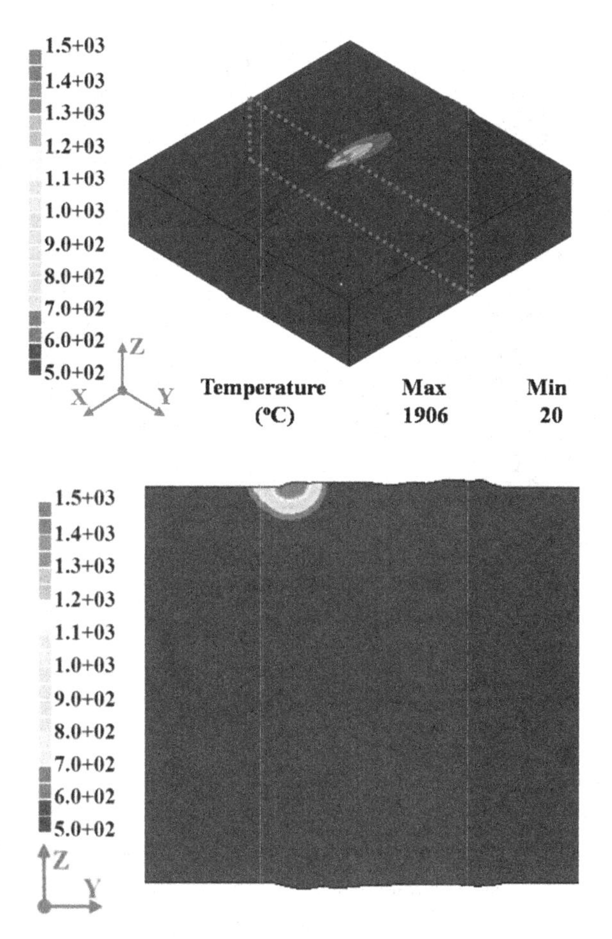

FIGURE 8.3 Computed transient temperature contour of examined butt-welded joint

Because of a large number of elements and welding lines for FE model of butt-welded joint with Q690 thick plate, parallel computation technology was employed to improve the computational efficiency of thermal analysis. With the high-performance computer of Dell X7910 with 48 cores, the time consumption with 8 cores of thermal analysis is 2 hours, 55 minutes, and 7 seconds (in total 10,507 seconds).

As the computed results of thermal analysis, transient temperature contour of butt-welded joint with Q690 thick plate and its middle cross-section was obtained as demonstrated in Figure 8.3, while welding arc is passing the middle region of welding line. It can be seen that maximal temperature is

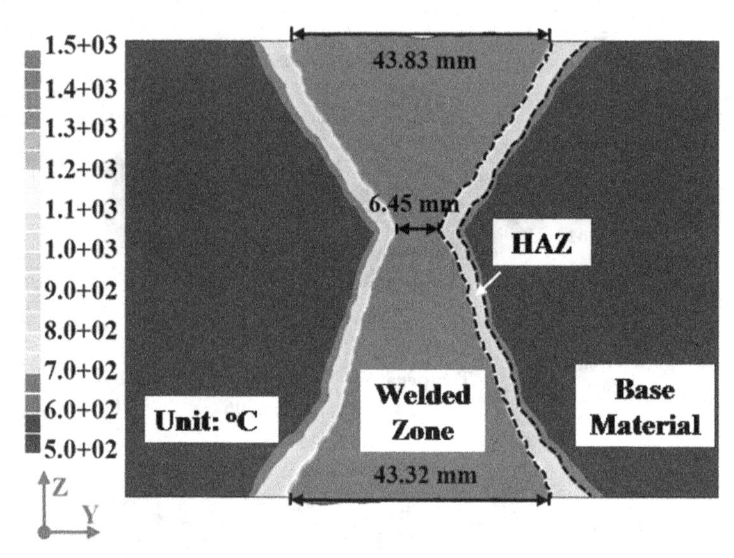

FIGURE 8.4 Computed maximal temperature contour of middle cross-section of butt-welded joint of Q690 thick plate

about 1900°C in the region of welded zone, and temperature above 1500°C to evaluate welded zone also can be obtained.

As shown in Figure 8.4, the maximal temperature contour of middle cross-section during entire welding process was demonstrated. Due to the multipass welding of examined butt-welded joint with Q690 thick plate, there are lots of thermal cycles for the points in the region of welded zone and heat-affected zone (HAZ), while complicated microstructure evolution and solid phase transformation occur. Besides, the widths of molten pool are 43.8 mm, 6.4 mm, and 43.3 mm in the location of top surface, root region, and bottom surface, respectively. Compared with measurement of molten pool of butt-welded joint of Q690 thick plate as shown in Figure 7.16, the differences in molten pool width are only 3.2%, 8.4%, and 4.8% for the welding top surface, welding root, and welding bottom surface, respectively.

In order to further examine computational accuracy of thermal analysis, Computed thermal cycle of corresponding location of experimental thermocouple 1 was compared with measured results as shown in Figure 8.5. There is good agreement between computed and measured thermal cycles not only for tendency but also for magnitude of maximal temperature. Therefore, with the comparison of molten pool geometry

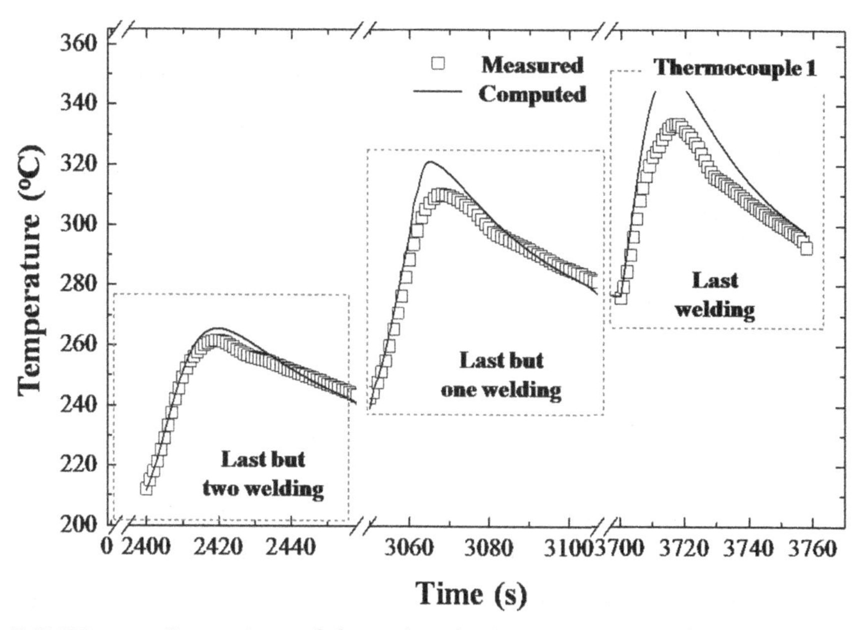

FIGURE 8.5 Comparison of thermal cycles between computed and measured results of butt-welded joint of Q690 thick plate

and thermal cycle to measurement, the computational accuracy of thermal analyses of butt-welded joint of Q690 thick plate was confirmed.

8.3 MICROSTRUCTURE EVOLUTION DURING WELDING

Based on the literature and research progress, the grain size of initial austenite is 7.6 μm, which will be used to predict the microstructure and hardness performance during the butt welding of Q690 thick plate. With the previous computed thermal cycle, chemical component of material as well as grain size of initial Austenite, some critical temperatures during solid transformation can be calculated as summarized in Table 8.1, while there is a little bit of difference between molten pool and HAZ due to the chemical components of materials.

By measurement of phase distribution in the region of molten pool and HAZ, volume fraction of martensite, Bainite, and Austenite as well as Ferrite and Perlite can be obtained, and their computed volume fractions were compared as summarized in Table 8.2 as Figure 8.6. Good agreement between computed and measured volume fraction of microstructure can be observed, while the difference can result from measurement tolerance as well as computed thermal cycles. Thus, the computed accuracy and

TABLE 8.1 Critical temperature during solid transformation due to welding.

Critical Temperature	Melton Pool (°C)	Heat-Affected Zone (°C)
Lower Critical Temperature (A1)	662	710
Upper Critical Temperature (A3)	755	778
Precipitate Dissolution Temperature (TS)	1,221	1,077
Liquids temperature (TL)	1,505	1,516
Solids temperature (TS)	1,472	1,497
Bainite Phase Transformation Temperature (BS)	467	517
Martensite Phase Transformation Temperature (MS)	291	383

TABLE 8.2 Comparison of computed and measured volume fraction of microstructure.

	Molten Pool		Heat-Affected Zone	
Microstructure	**Measurement**	**Computation**	**Measurement**	**Computation**
Martensite	0.300	0.223	0.806	0.808
Bainite	0.560	0.640	0.183	0.177
Austenite-Ferrite-Pearlite	0.140	0.137	0.011	0.015

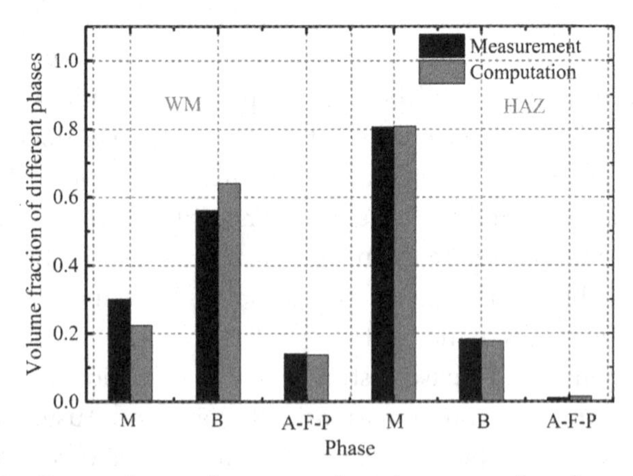

FIGURE 8.6 Comparison of computed and measured volume fraction of microstructure

modeling of microstructure evolution were both validated with the engineering acceptance.

With the FE model of butt-welded joint of Q690 thick plate as shown in Figure 8.1, Middle cross-section was selected to evaluate distribution of microstructure volume as demonstrated in Figure 8.7, while there are

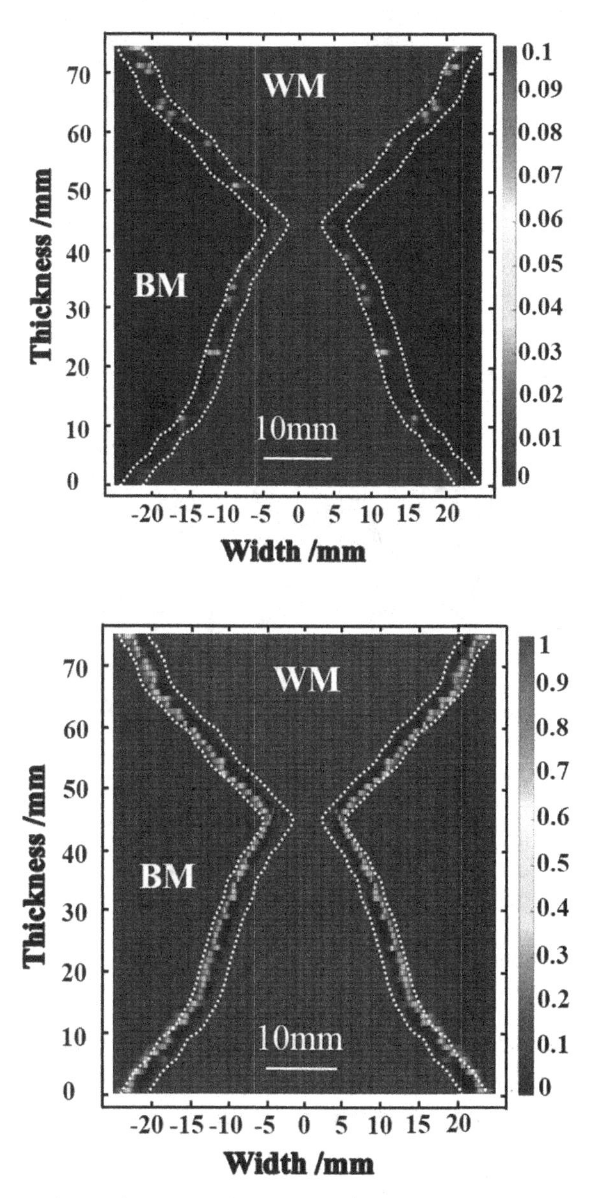

FIGURE 8.7 Volume fraction of microstructure of butt-welded joint of Q690 thick plate

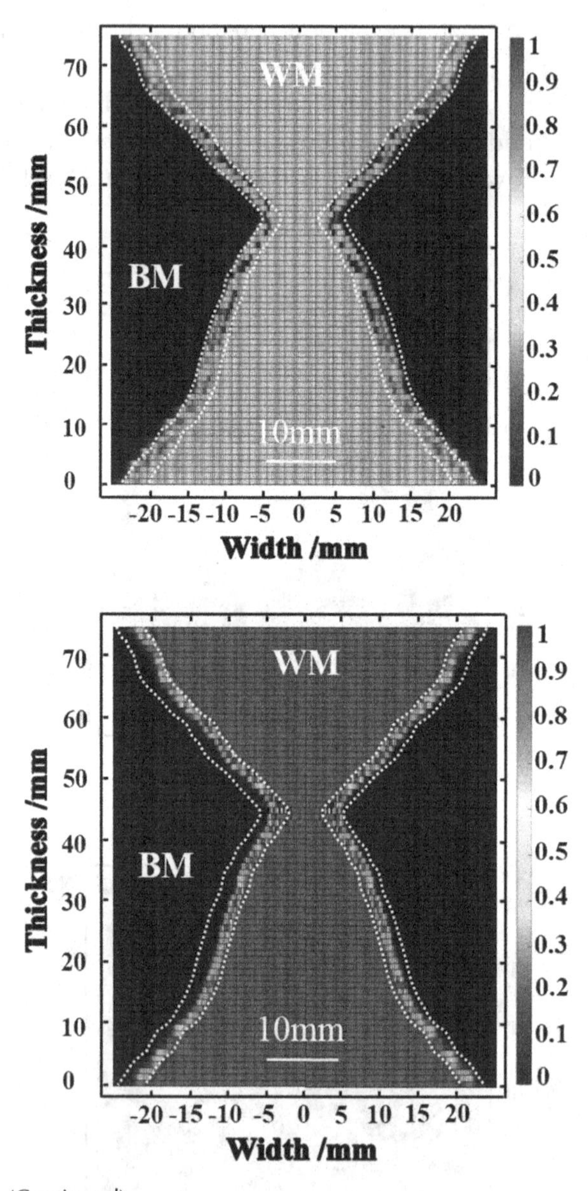

FIGURE 8.7 (Continued)

2910 nodes and each computation of microstructure evolution is about 120 seconds. It may be seen that there is less Austenite in the region of molten pool and HAZ due to the Austenite decomposition during cooling process.

In detail, the dominant microstructure in the region of base material is Ferrite and Pearlite. And there is a microstructure mixture of Bainite and Martensite as well as Ferrite and Pearlite in the region of HAZ. Moreover, hardness performance of butt-welded joint of Q690 thick plate can then be evaluated with the microstructure volume fraction.

In general, Austenite will sequentially decompose during the cooling process to be Ferrite, Pearlite, Bainite, and Martensite due to individual critical temperature, and hardness will be higher and higher with this sequence of microstructure decomposition. Refinement of grain size of Ferrite will improve the plasticity and toughness of material, and Martensite will enhance the strength and hardness of material due to its microstructure feature.

With the calculation formula of hardness as introduced in Eq. (2.50), total hardness of each node in the FE model of butt-welded joint of Q690 thick plate is the sum of hardness of Martensite, Bainite, Austenite, Ferrite, and Pearlite. In particular, hardness of individual microstructure will be determined not only by chemical components but also by cooling rate of each thermal cycle. Thus, the hardness evaluation of butt-welded joint of Q690 thick plate can be recognized with difficult and complex features.

In order to quantificational examine the volume fraction of microstructure of butt-welded joint of Q690 thick plate, points on line 1, line 2, and line 3 as demonstrated in Figure 8.8 were considered. Volume fraction of microstructure of points on considered lines was then presented as shown in Figure 8.9.

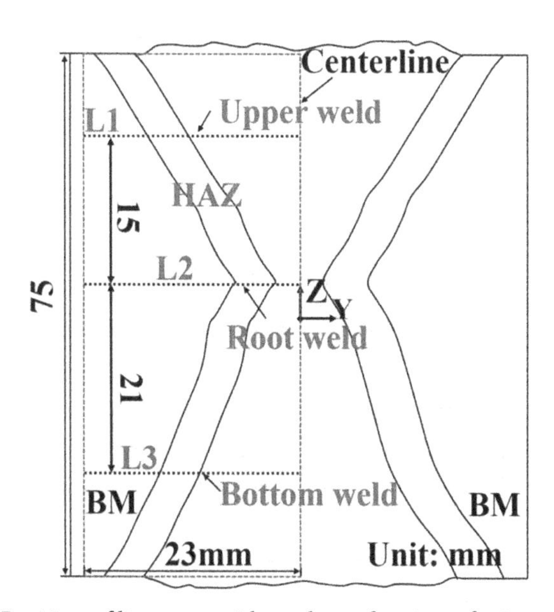

FIGURE 8.8 Position of line to consider volume fraction of microstructure

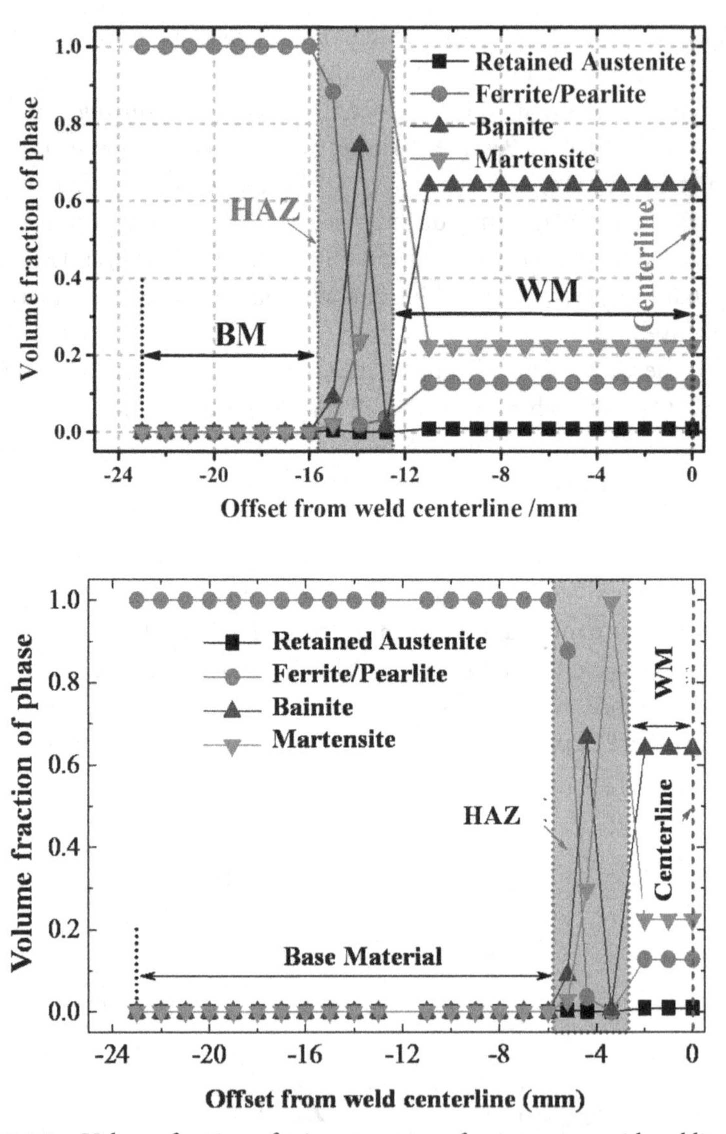

FIGURE 8.9 Volume fraction of microstructure of points on considered lines

It can be seen that there is a mixture of Ferrite and Pearlite in the region of base material. For the molten pool, the volume fractions of Bainite, Martensite, Ferrite/Pearlite, and remaining Austenite are 64%, 22.32%, 12.76%, and 0.92%, respectively. And there will be much more complex volume fraction of microstructure for the region of HAZ due to the complicated thermal cycle as well as solid phase transformation during welding.

Since volume fraction of microstructure is mainly determined by grain size of Austenite, maximal temperature, number of thermal cycles as well as cooling rate, microstructure evolution as well as solid phase transformation was then examined as demonstrated in Figure 8.10, while a typical point in the region of HAZ of butt-welded joint of Q690 thick plate was considered.

In addition, the maximal temperature and cooling rate are 970°C and 5.23°C/s, respectively. Due to the higher cooling rate, there is less time for Austenite to decompose to Ferrite and Pearlie, and the main decomposition is Bainite, which is the dominant type of microstructure.

Moreover, if there are lots of thermal cycles, Ferrite and Pearlite will transform to Austenite again during heating process and decompose during cooling process according to the later thermal cycle; in particular Bainite and Martensite will keep current condition without any solid phase transformation. Thus, the enhanced microstructure evolution model and in-house programming can be employed to evaluate volume fraction of microstructure and hardness performance of thick plate butt welding with multipass for scientific research and engineering application. Moreover, the optimized welding conditions can be proposed to modify the thermal

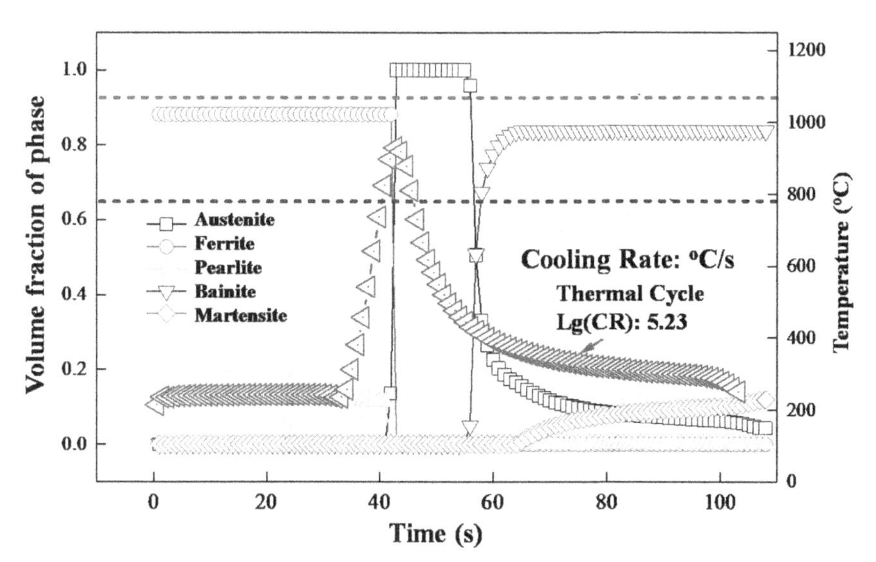

FIGURE 8.10 Microstructure evolution of typical point in the region of HAZ of butt-welded joint of Q690 thick plate

cycles as well as microstructure distribution, even hardness performance of butt-welded joint of Q690 thick plate.

With the above computation, volume fraction of microstructure of each point can be obtained, and Vickers hardness of points on middle cross-section will then be calculated based on volume fracture and hardness of different microstructures as demonstrated in Figure 8.11. It can be found that Vickers hardness in the region of base material and molten pool has an almost constant value, while higher Vickers hardness can be seen in the region of molten pool.

There is a variable value of Vickers hardness in the region of heat-affected zone. Microstructure in the region of molten pool is mainly Bainite, which will be Martensite in the region of heat-affected zone. Due to the higher Vickers hardness of Martensite compared with Bainite, the Vickers hardness of heat-affected zone is obviously higher than that of molten pool.

Taking the lines as indicated in Figure 8.8, Predicted Vickers hardness of points on examined lines were compared with the measurement to confirm the computational accuracy of proposed microstructure evolution model as demonstrated in Figure 8.12. There is a good agreement of Vickers hardness between computed and measured results of butt-welded joint with Q690 thick plate and the maximal tolerance is only 6%, which can be accepted for both research and application.

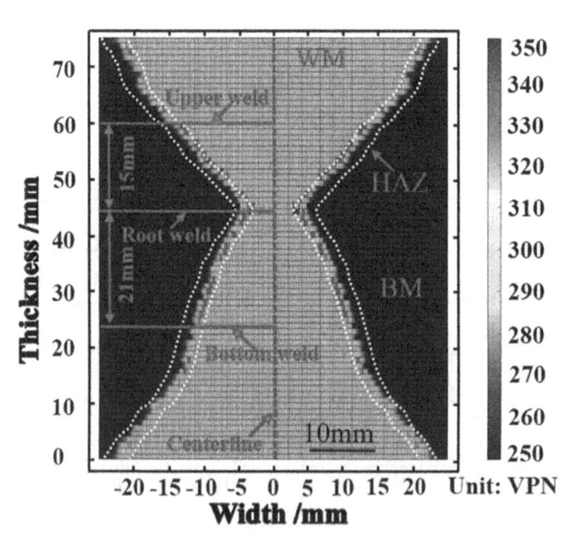

FIGURE 8.11 Computed hardness of points on middle cross-section of butt-welded joint with Q690 thick plate

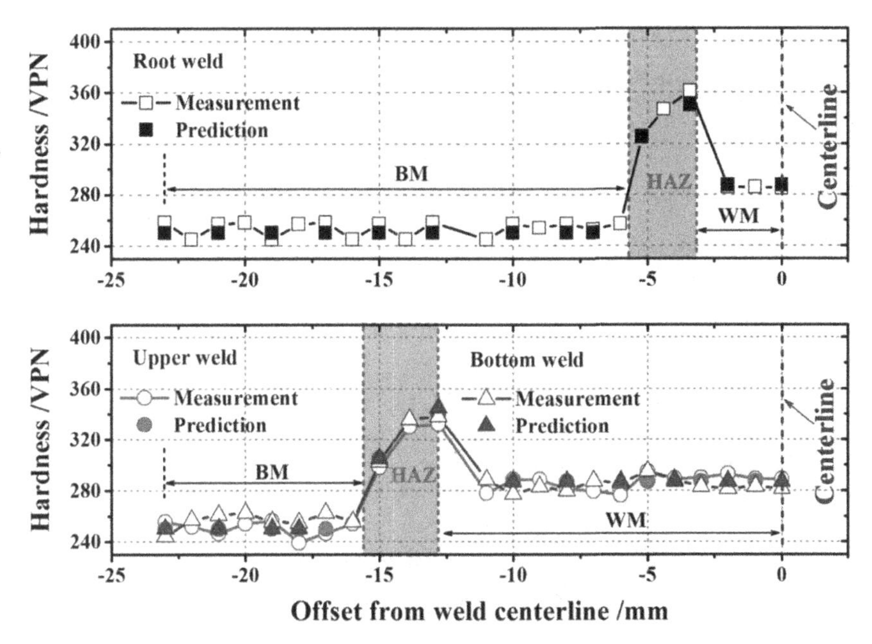

FIGURE 8.12 Comparison of Vickers hardness between computed and measured results of butt-welded joint with Q690 thick plate

8.4 PREDICTION OF LONGITUDINAL RESIDUAL STRESS

Since the volume variation due to solid phase transformation as well as initial stress caused by TMCP has significant influence on welding residual stress, there will be four computational cases as summarized in Table 8.3.

Based on the in-house program for welding mechanical analysis, FE model as shown in Figure 8.1 and boundary condition to prevent rigid body motion as indicated in Figure 8.13 were considered, while previous computed temperature was taken as the thermal loading.

Welding residual stress of butt-welded joint with Q690 thick plate was examined by means of FE analysis, while parallel computation technology was employed to enhance computational efficiency. In detail, the time consumption with 6 cores of mechanical analysis is 20 days 16 hours 44 minutes 24 seconds (in total 1,788,264 seconds), which is a massive requirement of computation for engineering applications.

After computation, welding residual stress was modified by considering the volume variation due to solid phase transformation and rolling initial stress of each point of butt-welded joint with Q690 thick plate, and predicted result of welding residual stress then could be obtained.

TABLE 8.3 Welding residual stress computational cases.

Computational Case No.	Description
Case 1	Without solid phase transformation and rolling initial stress
Case 2	Considering rolling initial stress
Case 3	Considering solid phase transformation
Case 4	Considering both solid phase transformation and rolling initial stress

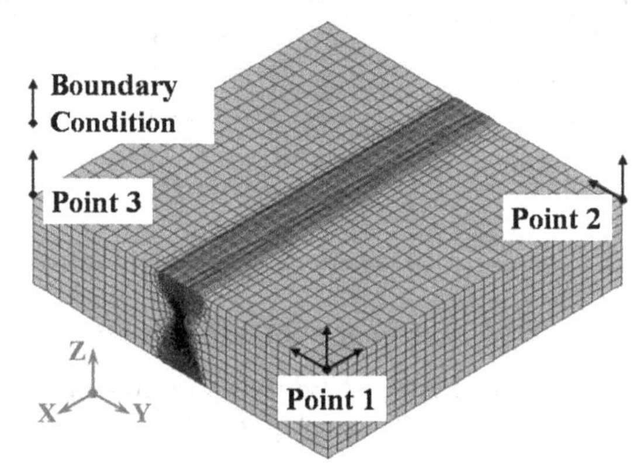

FIGURE 8.13 Mechanical boundary condition of examined butt-welded joint of Q690 thick plate

8.4.1 Influence of Solid Phase Transformation

There have been some research works about solid phase transformation during welding by means of conventional thermal elastic plastic FE computation, while influence of only one phase transformation such as Martensite transformation on welding residual stress was considered [117–119].

With the previous investigation, there will be four phase transformations for Austenite decomposition during cooling process, which should all be considered to enhance the computational accuracy. In detail, for the austenization of the initial microstructure during the heating process, BCC crystal of Ferrite/Pearlite will be transferred to be FCC crystal of Austenite with volume shrinkage.

Meanwhile, during the cooling process, Austenite will decompose and BCC crystal of microstructure such as Ferrite, Pearlite, Bainite, and Martensite with volume expansion. Due to the volume expansion during cooling process, welding residual stress will be relaxed and modified.

Volume variation and its generated phase transformation strain is usually considered with plastic strain to be total residual strain for welding residual stress evaluation [117–119]. Based on the total volume of microstructure while considering solid phase transformation, volume variation was employed and proposed to examine the influence of phase transformation on welding residual stress. When the volume fractions of each microstructure are known, the total volumes of examined points before and after solid phase transformation can be calculated.

While the volumes of Austenite (V_{RA}), Ferrite (V_F), Pearlite (V_P), Bainite (V_B), and Martensite (V_M) are 8 μm^3, 12 μm^3, 12 μm^3, 32 μm^3, and 32 μm^3, respectively [139–143]. Therefore, total volumes of microstructure before and after phase transformation can be obtained with Eq. (8.1) and Eq. (8.2).

$$V_{before} = X_F \times V_F + X_P \times V_P \tag{8.1}$$

$$V_{after} = X_M \times V_M + X_B \times V_B + X_F \times V_F + X_P \times V_P + X_{RA} \times V_{RA} \tag{8.2}$$

where X_F, X_P, X_M, X_B, and X_{RA} mean the volume fraction of Ferrite (F), Pearlite (P), Martensite (M), Bainite (B), and remaining Austenite (A).

8.4.2 Influence of Initial Stress

For high tensile strength steel such as Q690, the initial stress has some features, given as follows [102, 106, 120]:

(1) Rolling initial stress will be self-balance in the thickness direction with symmetrical distribution.

(2) Rolling initial stress will be compressive in the top and bottom surfaces, and tension in the middle region of thickness direction.

(3) Maximal magnitude of rolling initial compressive stress in the longitudinal and transverse directions will be about 30–40% of yield stress of high tensile strength steel

Besides, initial transverse and longitudinal stresses have almost identical distribution in the plate thickness direction, and based on the relationship between stress ratio of high tensile strength steel which is usually defined as initial stress/yield stress and plate thickness, the initial stress of Q690 thick plate can be solved. Therefore, initial transverse and longitudinal stress has almost identical distribution of Q690 thick plate were obtained as shown in Figure 8.14.

For the butt-welded joint of Q690 thick plate, welding line will expand and shrink during the heating and cooling processes; however, due to the self-constraint of surrounding base material, residual stress will be generated, which also be influenced by volume variation due to microstructure transformation during cooling.

Based on the critical temperature of solid phase transformation, all nodes of FE model of examined butt-welded joint of Q690 thick plate could be divided into welding line ($T_{max} \geq TL$), heat-affected zone ($A1 < T_{max} < TL$), and base material ($T_{max} < A1$), which the welding residual stress will be different features for the abovementioned regions. In detail, welding line is generated by means of fusion and solidification of filler metal and partial base material, while only the influence of solid phase transformation due to welding should be considered. For the heat-affected zone, not only the solid phase transformation due to welding but also the initial rolling stress of base material should be considered to influence the welding residual stress. There will be no solid phase transformation due to welding in the region of base material, and only the initial rolling stress of base material should be considered.

Thus, welding residual stress of each node of FE model of examined butt-welded joint of Q690 thick plate can be obtained with Eq. (8.3) as follows:

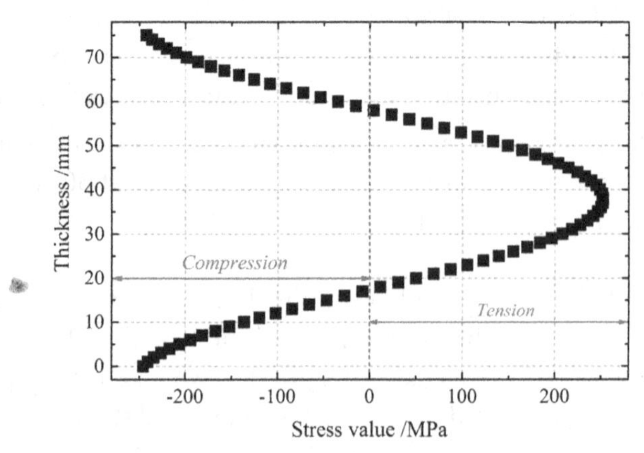

FIGURE 8.14 Distribution of initial stress in the thickness direction of Q690

$$\sigma_{after} = \begin{cases} \dfrac{\sigma_{before} V_{before}}{V_{after}}, \; T_{max} \geq T_L \\ \dfrac{\sigma_{before} V_{before}}{V_{after}} + \sigma_{initial}, \; A_1 < T_{max} < T_L \\ \sigma_{before} + \sigma_{initial}, \; T_{max} < A_1 \end{cases} \tag{8.3}$$

where σ_{before} and σ_{after} mean the welding residual stress of butt-welded joint of Q690 thick plate without and with considering the influence, respectively, MPa; $\sigma_{initial}$ means the initial rolling stress, MPa; T_{max} means the maximal temperature during heating and cooling processes, °C; and A1 and TL are the lower critical temperature and liquid temperature, °C.

As shown in Figure 8.15, computed longitudinal and transverse residual stresses were demonstrated considering the solid phase transformation and rolling initial stress as computational case 4.

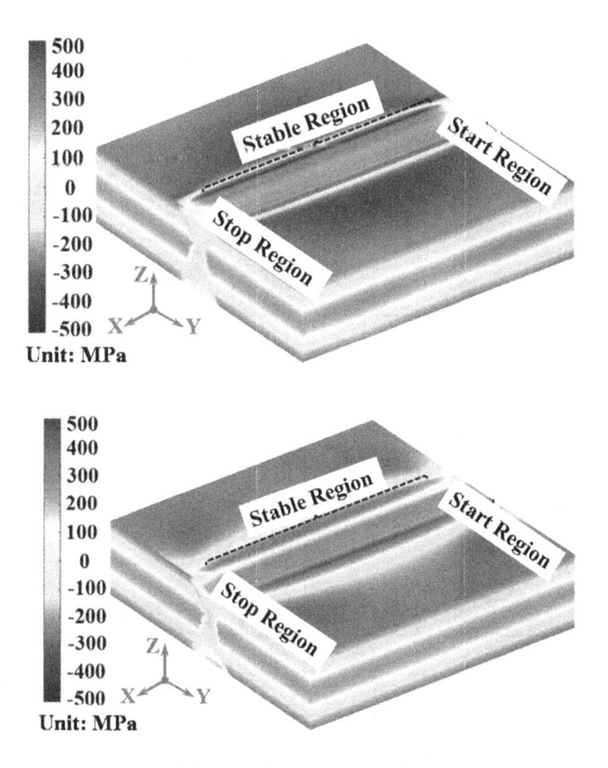

FIGURE 8.15 Computed welding residual stress of examined butt-welded joint of Q690 thick plate considering both solid phase transformation and rolling initial stress

8.4.3 Comparison with Measurement

Figure 8.16 shows the contour of longitudinal residual stress distribution of four simulation cases, respectively. In Case A, both SSPT and initial stress are utterly neglected. Tensile stress is mainly in the weld vicinity, while compressive stress is in root weld and BM. In Case B, distribution of longitudinal residual stress in BM and HAZ is changed when considering the effect of initial stress. Compressive stress can be seen near the plate surface and high tensile stress can be found at the midthickness weld. Moreover, initial stress can decrease the residual stress value in HAZ. In Case C, residual stress value in WM and HAZ is obviously reduced when taking the SSPT into account. In Case D, distribution shape of longitudinal residual stress in butt-welded joint is completely changed when considering the effect of SSPT and initial stress. Overall, a similar trend of residual stress in WM can be seen from the results except that tensile stress value in Case C and Case D is obviously lower than that in Case A and Case B. Peak value of longitudinal residual stress in the WM had a good agreement compared with

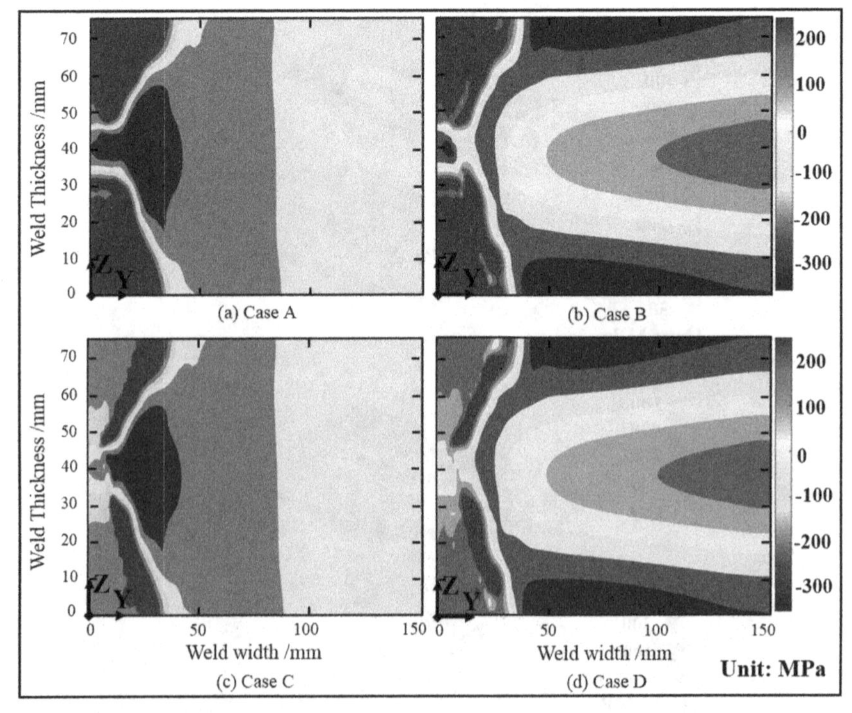

FIGURE 8.16 Computed distribution of longitudinal residual stress of examined butt-welded joint with Q690 thick plate

previous result. Results suggested that SSPT and initial stress play a significant role in the formation of longitudinal residual stress during multipass butt welding of Q690 HSS thick plate.

In Case D, high tensile stress value can be found in the area adjacent to HAZ, which will be potentially harmful for the fracture property. Obvious compressive stress can be seen in the root weld. On the one hand, formation of compressive stress is due to the heat treatment of latter welds. On the other hand, when carrying out the filling welding of thick weld, the melted weld metal is constrained by the root weld, resulting in the generation of compression stress. Figure 8.17 displays the comparisons of longitudinal residual stress profile between experimental and simulated results on two through-thickness lines (at 0 and 5 mm from the weld centerline). At the weld centerline, only the effect of SSPT is taken into consideration. At 5 mm away from the weld centerline, besides SSPT the effect of initial stress has to be considered since it can change the residual stress value at the midthickness weld. Volume expansion caused by SSPT during the cooling process can suppress the shrinkage of WM, thereby decreasing the maximal tensile stress value. Peak value of tensile stress in Case D is obviously lower than yield stress of Q690 HSS, which occurs at a depth of 5 mm.

Significant deviations between the experimental and numerical datasets remain, particularly near plate surface and at midthickness weld. From the results of Case D shown in Figure 8.17, compressive stress distributes at the root weld, while it is tensile stress obtained by CM. According to the previous works [144–146], tensile stress can be found at the root weld when measuring the residual stress distribution in thick welded joint by CM.

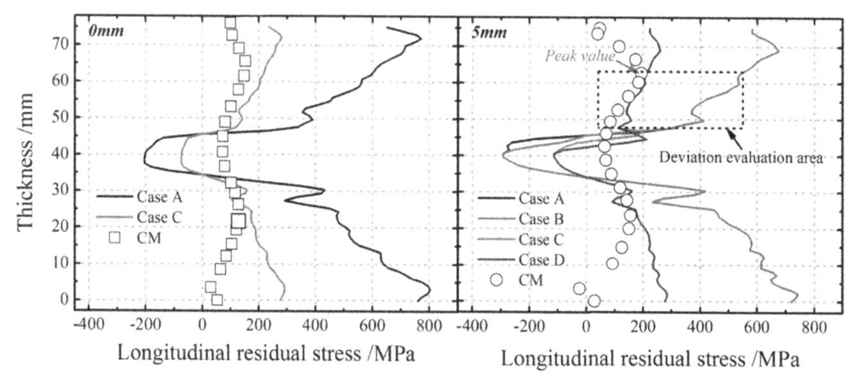

FIGURE 8.17 Distribution of longitudinal residual stress in thickness direction of points away from welding line (0 mm and 5 mm)

What's more, error of stress caused by edge effect is inevitable, so compressive stress can be seen in the weld near the plate surface. When predicting welding residual stress, longitudinal residual stress value near plate surface is obviously larger than other regions. In this study, some data are taken for error analysis, as shown in Figure 8.18. The RMSE is 8 for the deviation evaluation area when considering the effect of SSPT, while the RMSE is 307 for the deviation area when the effect of SSPT is not considered. However, obvious deviation of residual stress between simulation and measurement at HAZ and the surface of WM may be attributed to the aging effect which is neglected in the numerical simulation methods. Overall, residual stress distribution by numerical methods is in reasonable agreement compared with the measurement. Figure 8.17 shows the comparisons of longitudinal residual stress profile between experimental and simulated results on four through-thickness lines (at 25, 50, 75, and 100 mm from the weld centerline). Residual stress profile, at 25 mm from weld centerline, exhibited a "W" shape with residual stress in tension near the plate surface

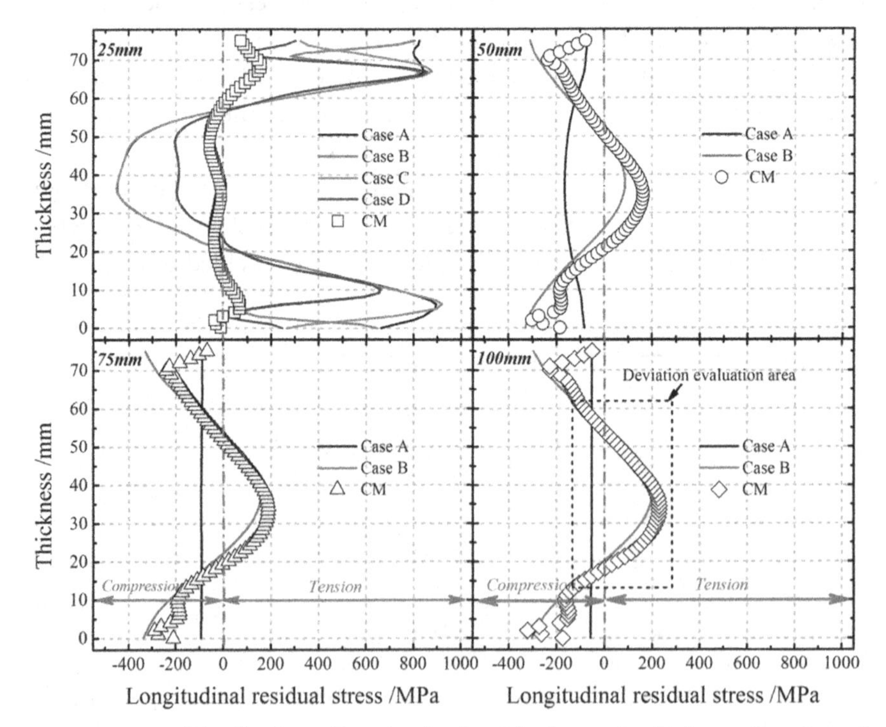

FIGURE 8.18 Distribution of longitudinal residual stress in thickness direction of points away from welding line (25 mm, 50 mm, 75 mm, and 100 mm)

balanced with compression at depths of about 10 and 60 mm. For the other three lines, the distribution of residual stress between Case B and CM was extremely consistent. As can be seen, initial stress in BM can turn compressive stress at middle section into tensile stress and increase compressive stress value near plate surface. Moreover, residual stress distribution in BM exhibited a "V" shape with the residual stress in compression up to 300 MPa near the surface balanced with tension (~200 MPa) at depths of 20 and 50 mm. In addition, the RMSE in BM is 11 for the deviation evaluation area when considering the effect of initial stress, while the RMSE is 137 for the deviation area when the effect of initial stress is not taken into account.

To further analyze the individual effect on the formation of residual stress in HAZ, Figure 8.19 displays the residual stress distribution away from the weld centerline. When taking SSPT and initial stress into consideration (Case D), residual stress distribution matches well with the measurement. From Figure 8.19a, residual stress in HAZ is tensile stress when the SSPT and initial stress are utterly neglected. In Case B, stress value near the BM (region 2) is obviously reduced due to the initial stress, and in Case C stress value near the fusion line (region 1) is significantly reduced due to the SSPT. In Figure 8.19b, residual stress in HAZ is compressive stress when the SSPT and initial stress are not taken into account. Stress value (region 3 and region 4) is obviously reduced when considering the individual effect of SSPT (Case C) and initial stress (Case B), while in Case D, SSPT decreases the compressive stress value near the BM (region 4), and then initial stress turns the compressive stress into tensile stress. Therefore, it can be concluded that SSPT has a significant role in the magnitude of residual stress in HAZ near the WM and initial stress determines the residual stress value in HAZ near the BM.

8.5 EVALUATION OF TRANSVERSE RESIDUAL STRESS

Compared with the experimental data obtained by CM, the proposed numerical methods are verified to be accurate and reliable. Therefore, the distribution of transverse residual stress in butt-welded joint of Q690 HSS was further estimated in this paper. Figure 8.20 shows the contour of transverse residual stress distribution of four simulation cases, respectively. As can be seen from the figures, distribution shape of transverse residual stress profile in butt-welded joint is also completely changed when considering the effect of SSPT and initial stress. In Case A, high tensile stress in the weld zone is mainly near the surface and compressive stress occurs at the root

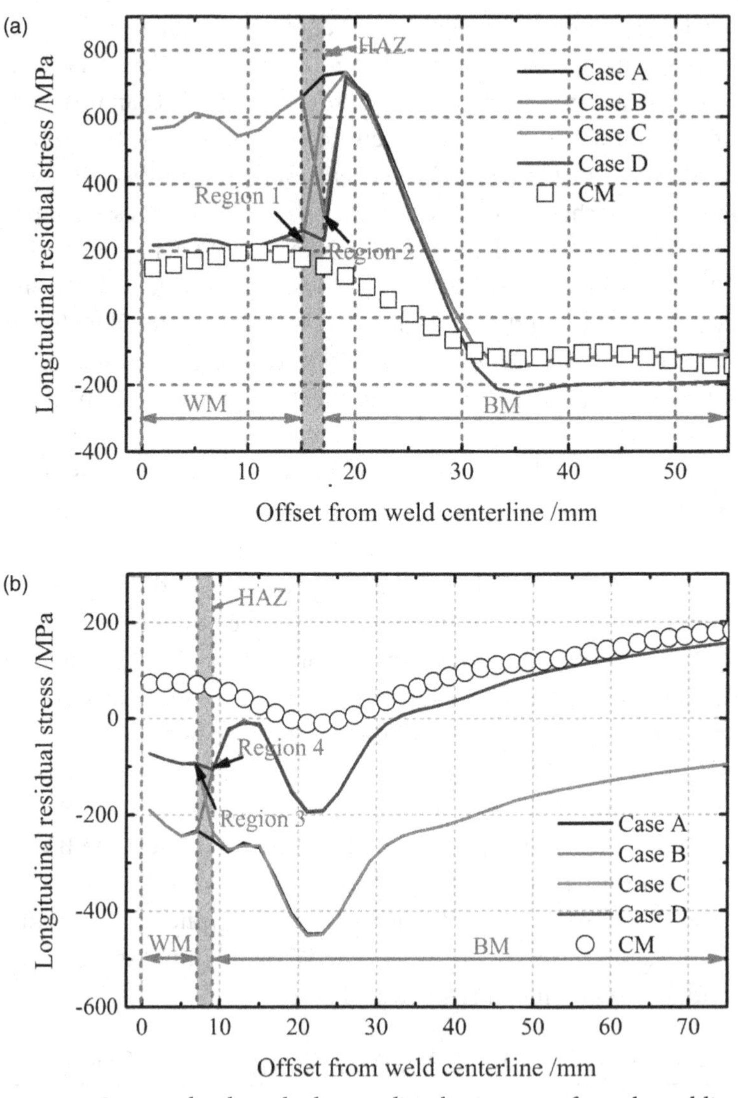

FIGURE 8.19 Longitudinal residual stress distribution away from the welding line

weld. In Case C and Case D, tensile stress value in the weld zone is obviously reduced due to the SSPT. The maximal value of tensile stress in the area adjacent to HAZ is about 10 mm from the surface and peak value of compressive stress is at the midthickness plate. Moreover, it can be observed from Case B and Case D that compressive stress in BM is near the surface and tensile stress is at the middle section of BM. Figure 8.21 shows the comparison of transverse residual stress profile between Case A and Case

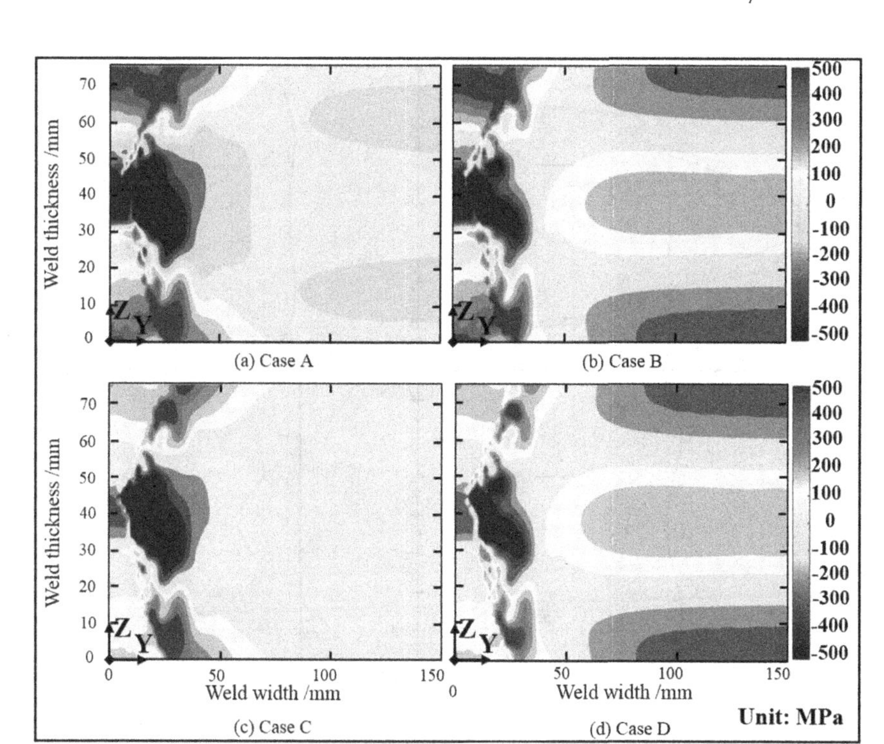

FIGURE 8.20 Contour of transverse residual stress of examined butt-welded joint of Q690 thick plate with different cases

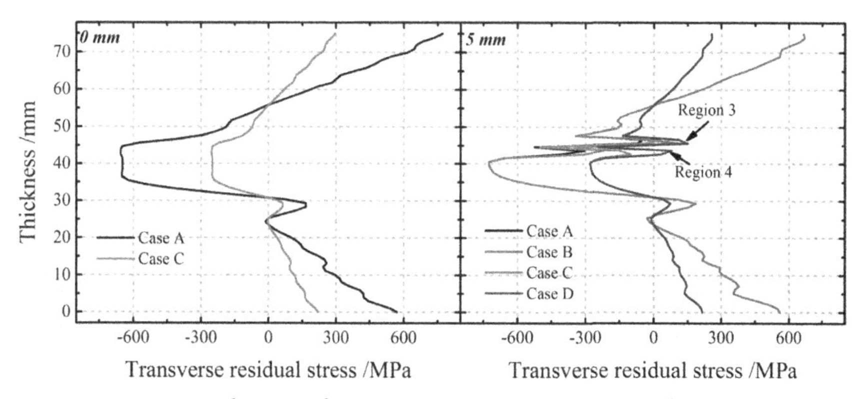

FIGURE 8.21 Distribution of transverse residual stress in thickness direction of points away from welding line (0 mm and 5 mm)

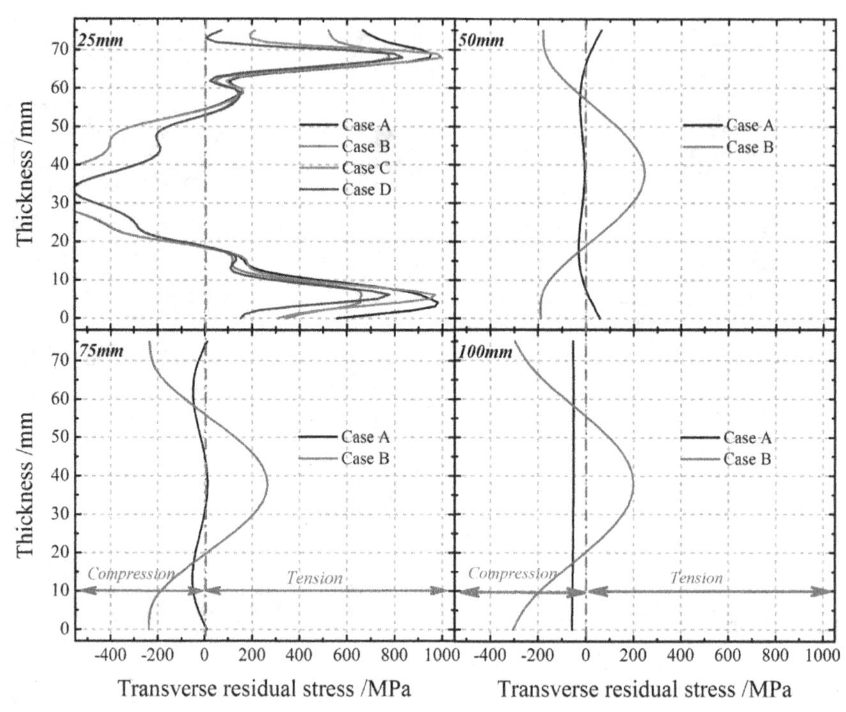

FIGURE 8.22 Distribution of transverse residual stress in thickness direction of points away from welding line (25 mm, 50 mm, 75 mm, and 100 mm)

B on two through-thickness lines (at 0 and 5 mm from the weld centerline). When considering the effect of SSPT, distribution law of transverse residual stress at the weld centerline is similar but the residual value is significantly reduced. However, transverse residual stress distribution, at 5 mm from the weld centerline, is obviously changed when considering the effect of SSPT and initial stress. Moreover, compressive stress in region 3 and region 4 is reduced due to the SSPT and then transformed into tensile stress due to the initial stress. Figure 8.22 shows the comparisons of transverse residual stress profile between Case A and Case B on four through-thickness lines (at 25, 50, 75, and 100 mm from the weld centerline). For line at 25 mm offsetting from the weld centerline, either SSPT or initial stress can reduce the tensile stress near plate surface. At midthickness plate, only initial stress can reduce the compressive stress. However, the distribution law of transverse residual stress was obviously changed at 50, 75, and 100 mm from the weld centerline. As can be seen, initial stress turned tensile stress near the surface into compressive stress and transformed compressive stress at midthickness BM into tensile stress.

8.6 SUMMARY

Welding residual stress of butt-welded joint with Q690 thick plate was evaluated by means of transient thermal elastic plastic FE computation, while solid element FE model was employed. Computed result of thermal analysis was validated by measured thermal cycles as well as shape of molten pool. In order to accurately evaluate welding residual stress, microstructure evolution as well as solid phase transformation was considered and validated with measurement. Eventually, FE computation considering solid phase transformation and initial rolling stress was carried out to predict welding residual stress, which has a good agreement compared with measurement.

CHAPTER 9

Fracture of Welded Joint with GTN Model

WELDED JOINT IS THE critical part of the complex and large welded structure; the failure or damage under the external loading will usually begin from the welded joint. The elementary cause of welding process to generate mechanical failure can be understood by welding defects as well as welding residual stress, which could be represented by FE computation with GTN model.

9.1 TENSILE EXPERIMENTS OF MATERIAL

With the axial tensile test, elementary mechanical performances of base material, welding filler, and butt-welded joint can be obtained, which includes elastic modulus, yield stress, ultimate stress, fracture strength, and elongation rate. Universal testing machine with MTS E45.305 as shown in Figure 9.1 was employed to carry out the tensile test, while the maximal tensile load is about 200 KN.

Based on the test benchmark, geometrical profile and dimension of axial tensile test specimen were designed, and axial tensile test specimen will be fabricated by means of wire-electrode cutting with spark erosion, while the position and dimension of axial tensile test specimen were demonstrated in Figure 9.2. In addition, due to the actual situation, axial tensile test

DOI: 10.1201/9781003442523-9

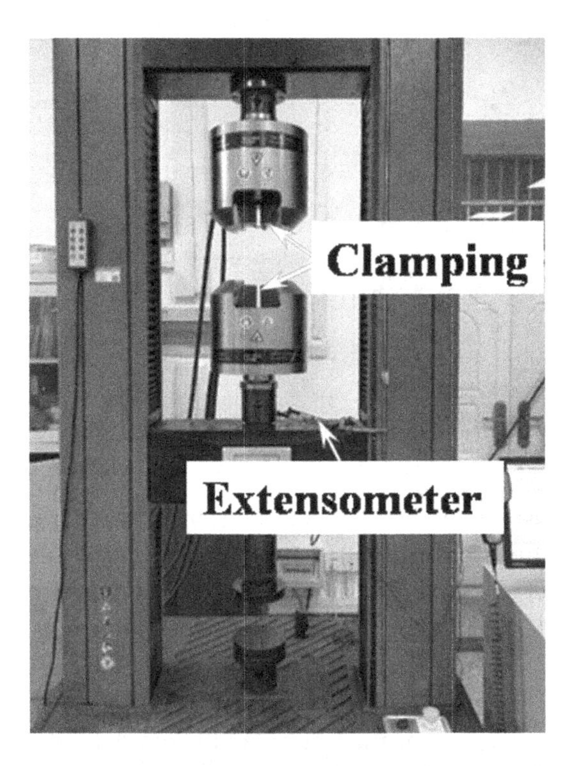

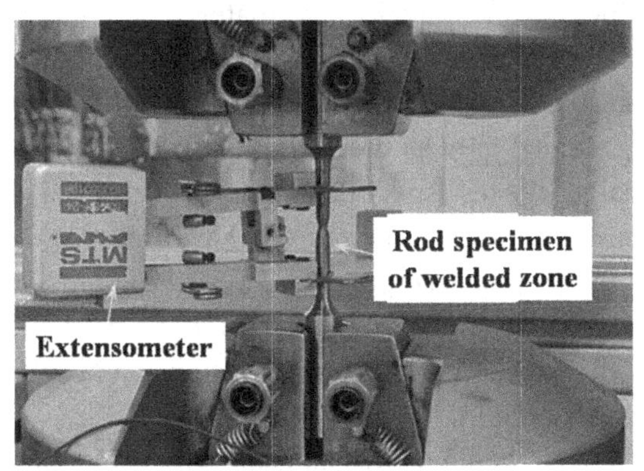

FIGURE 9.1 Universal testing machine for axial tensile test

specimen of base material and butt-welded joint was designed as rectangle specimen, and axial tensile test specimen of filler metal is rod specimen. Moreover, gauge length of extensometer for rectangle specimen is 50 mm, and gauge length of extensometer for rod specimen is 25 mm, while the speed of axial tensile test is about 5 mm/min.

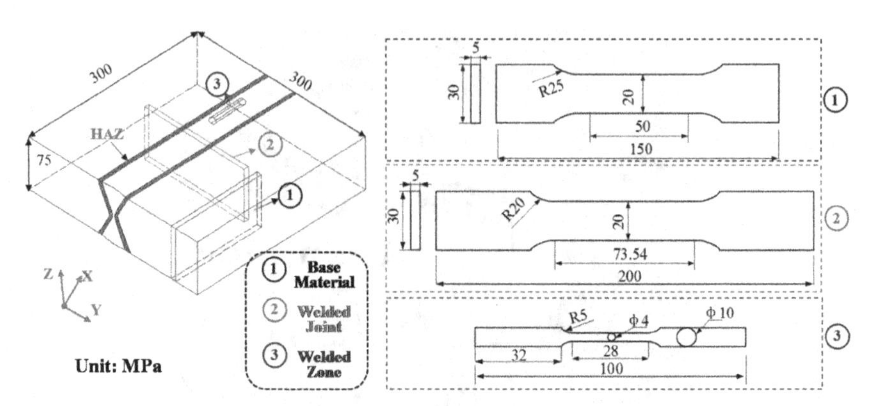

FIGURE 9.2 Position and dimension of axial tensile test specimen

In order to avoid the test tolerance, there are all three specimens of axial tensile test for the base material, filler metal, and butt-welded joint.

With the average result of axial tensile measured data, stress-strain curve of the base material, filler metal, and butt-welded joint were compared and demonstrated in Figure 9.3. It could be seen from the engineering stress-strain curve as indicated in Figure 9.3(a) that filler metal has the highest ultimate strength and elongation rate, and the mechanical performance of butt-welded joint is determined by base material and filler metal, together. For the computation of axial tensile test, the engineering stress-strain curve should be converted to true stress-strain curve of the base material, filler metal, and butt-welded joint as shown in Figure 9.3(b).

After the axial tensile test, specimen of the base material, filler metal, and butt-welded joint will fail with fracture behavior as shown in Figure 9.4. It also may be found that fracture position of axial tensile test specimen of butt-welded joint is located at the region of base materials, which could confirm the excellent fracture performance and welding quality of examined butt-welded joint.

In addition, the elementary mechanical performances of base material, welding filler, and butt-welded joint were summarized in Table 9.1, and their parameters of strain hardening were also summarized in Table 9.1. It also could be seen that they have almost identical values of mechanical performance due to equal strong match for engineering application. Table 9.2 shows mechanical performances of base material, welding filler as well as butt-welded joint.

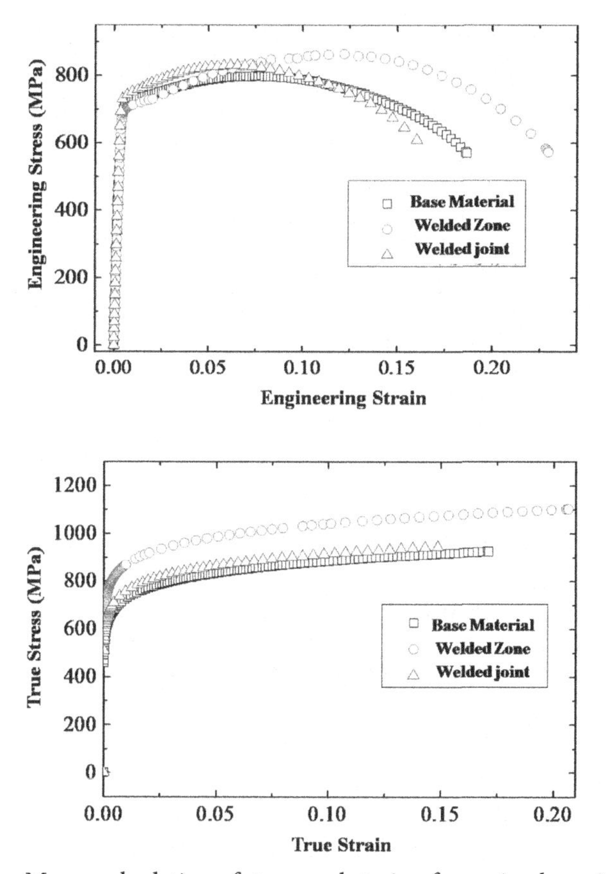

FIGURE 9.3 Measured relation of stress and strain of examined specimen

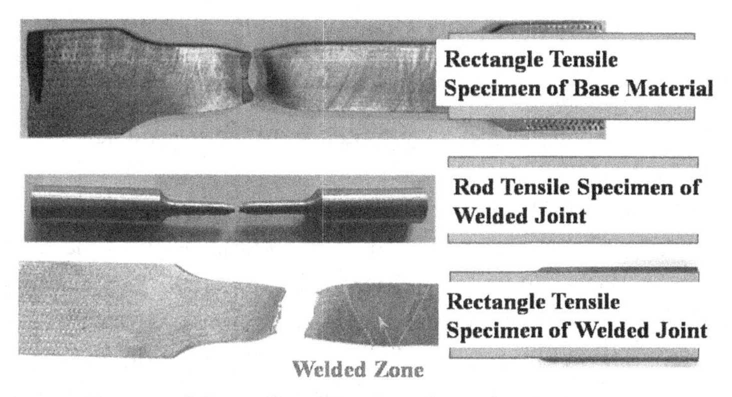

FIGURE 9.4 Fracture failure of examined axial tensile test specimen

TABLE 9.1 Mechanical properties of examined Q690 welded joint.

	Elastic Modules (MPa)	Yield Strength (MPa)	Ultimate Strength (MPa)	Fracture Strength (MPa)	Elongation
Base Material	210,000	710.1	799.2	569.7	0.19
Filler Metal	192,000	702.0	863.7	572.0	0.23
Butt-Welded Joint	210,000	707.6	829.9	609.4	0.16

TABLE 9.2 Parameters of strain hardening of base material, welding filler, and butt-welded joint.

	Strain Hardening Efficiency	Strain Hardening Index
Base Material	1 074 MPa	0.084
Filler Metal	1 247 MPa	0.078
Butt-Welded Joint	1 103 MPa	0.081

9.2 FRACTURE MODELING BASED ON GTN MODEL

With the GTN damage model to examine the fracture behavior of butt-welded joint of Q690 thick plate, parameters of GTN damage model should be confirmed beforehand. As introduced in the previous section, there will be nine parameters of GTN damage model, which are q_1, q_2, q_3, f0, f critical, f fracture, ε nucleation, S nucleation, and f nucleation.

As summarized in Table 9.2, parameters of GTN damage model for different high tensile strength steel were introduced [126–128]. In addition, q_1 and q_2 are usually in the range of 1.2–1.5 and 0.8–1.0, respectively. The ranges of ε nucleation and S nucleation were 0.1–0.65 and 0.005–0.1, respectively. For the different high tensile strength steels with individual yield stress, the situation that $q_1 = 1.5$, $q_2 = 1.0$, ε nucleation = 0.3, and S nucleation = 0.1 will yield in a computed result with good agreement with measured data [123, 124].

In general, initial void volume fraction (f0) is determined by microdefects of base materials such as microflaw, nonmetallic inclusion, and air hole. There will be fewer microdefects of Q690 base materials due to the treatment of strict thermal machinery rolling process, and it will be easy to generate welding defects of nonmetallic inclusion and air hole in the region of molten pool because of solidification of molten filler metal. Thus, the initial void volume fraction (f0) of molten pool will be obviously higher than that of base material.

Besides, there will be complex distribution of microstructure in the region of heat-affected zone; however, initial void volume fraction (f0) of HAZ can be considered to be that of base material. Due to the difference in microdefects, parameters of GTN damage model of butt-welded joint of Q690 thick plate are determined by that of base material and filler metal together. Moreover, base material and filler metal are both C-Mn low-carbon alloy steel, and the critical void volume fraction (f critical) and fracture void volume fraction (f fracture) of base material and filler metal are 0.15 and 0.17, respectively.

Therefore, some parameters of GTN damage model of base material and filler metal could be confirmed as: $q_1 = 1.5$, $q_2 = 1.0$, $q_3 = 2.25$, $f\ critical = 0.15$, $f\ fracture = 0.17$, $\varepsilon\ nucleation = 0.3$, and $S\ nucleation = 0.1$. The other parameters of GTN damage model of base material and filler metal, such as initial void volume fraction (f0) and void nucleation particle volume fraction (f nucleation), should be further examined.

9.2.1 Parameters Investigation of GTN Model of Base Material

With the ultrasonic inspection, initial void volume fraction (f0) of base material of Q690 thick plate can be quantificationally confirmed, the value of which is 0.00025. For the void nucleation particle volume fraction (f nucleation) of base material of Q690 thick plate, axial tensile test and FE computation were both carried out to confirm its accurate value.

Based on the dimension of rectangle specimen of base material tensile test as indicated in Figure 9.2, FE model with solid element of C3D8R was made as demonstrated in Figure 9.5. Due to the tiny size of void, element size is usually considered to be 0.1–1 mm, which will be appropriate to

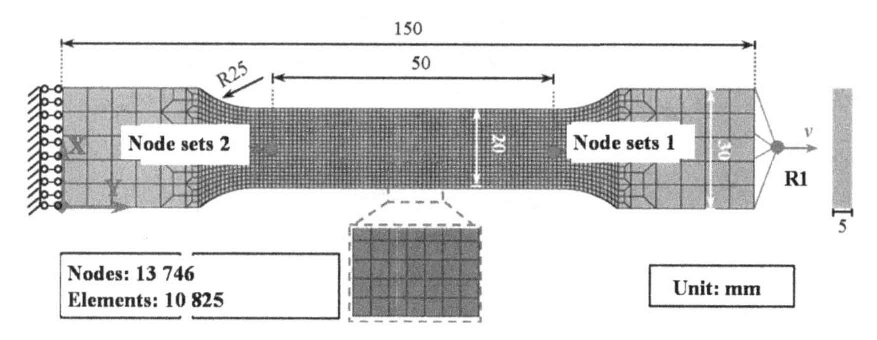

FIGURE 9.5 FE model of rectangle specimen of base material tensile test

actually represent evolution process of void nucleation, propagation, aggregation, and fracture of examined material specimen [126].

As shown in Figure 9.5 of FE model of rectangle specimen of base material tensile test, there are 13,746 nodes and 10,825 elements, while the element in the measured region is 1 mm in length, 1 mm in width, and 1 mm in thickness. Besides, the nodes in the clamping region of extensometer were selected to be node sets 1 and 2 as indicated in Figure 9.5. The FE model of rectangle specimen of base material tensile test was constrained to prevent the movement and rotation, and the tensile loading was applied to reference nodes 1. After FE computation, the results of tensile loading of reference nodes 1 as well as relative displacement of node sets 1 and 2 can be obtained, then the relationship between stress and strain of rectangle specimen of base material tensile test can be established as shown in Figure 9.6.

With the GTN damage model, the elastic modulus with 210 GPa, Poisson's ratio of 0.3, and density of 7860 kg/m^3 were considered. Explicit dynamic analysis was carried out to examine the tensile test process of rectangle specimen of base material while the tensile speed is about 5 mm/min.

There are three cases of FE computation with void nucleation particle volume fraction (f nucleation) of 0.01, 0.015, and 0.02. The computed engineering stress and strain curves of base material with different values of void nucleation particle volume fraction (f nucleation) were then compared with measurement as demonstrated in Figure 9.6. It can be seen

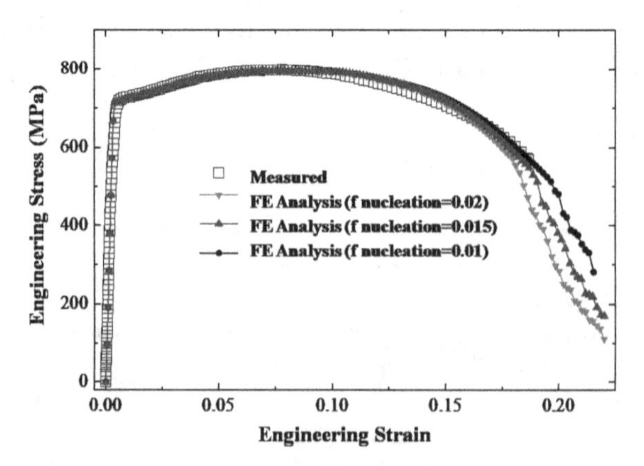

FIGURE 9.6 Comparison of engineering stress and strain curves of base material between computed and measured results

that there is a good agreement between computed and measured engineering stress and strain curves of base material. The computed results with different values of void nucleation particle volume fraction (f nucleation) were almost identical when the engineering strain is less than 0.1. When the engineering strain is bigger than 0.1, the lowering speed of engineering stress and strain curves of base material will increase with the increasing of void nucleation particle volume fraction (f nucleation).

With the definition of variance as introduced in Eq. (9.1), the influence of void nucleation particle volume fraction (f nucleation) on engineering stress and strain curve of rectangle specimen of base material was further examined. As shown in Figure 9.7, the variance will be lowest when the void nucleation particle volume fraction (f nucleation) equals 0.015. Thus, the appropriate value of void nucleation particle volume fraction (f nucleation) is 0.015 for the rectangle specimen of base material.

$$\text{Variance} = \sum_{i=1}^{n} \frac{\left(\sigma_{\text{computation}} - \sigma_{\text{measurement}}\right)^2}{n} \tag{9.1}$$

where n is the number of considered nodes; $\sigma_{computation}$ means computed engineering stress, MPa; and $\sigma_{measurement}$ means measured engineering stress, MPa.

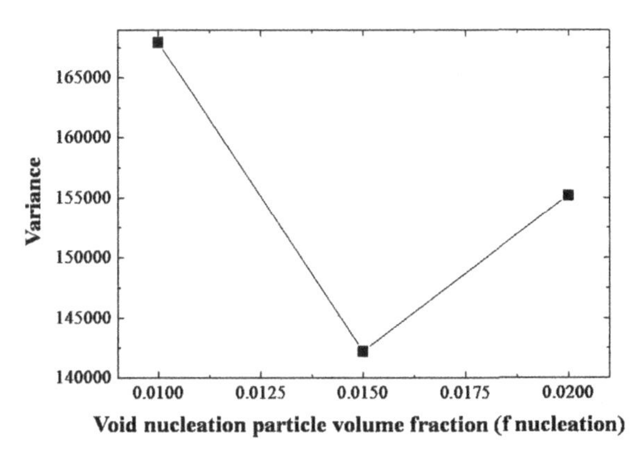

FIGURE 9.7 Variance with different values of void nucleation particle volume fraction (f nucleation) of rectangle specimen of base material

9.2.2 Parameters Investigation of GTN Model of Filler Metal

Based on the measurement as indicated in Figure 9.8, initial void volume fraction (f0) of filler metal of examined butt-welded joint can be quantificationally confirmed, which value is 0.01. Similar to previous tensile test and FE computation, void nucleation particle volume fraction (f nucleation) of welded zone of examined butt-welded joint was also obtained.

As mentioned above, rod specimen of welded zone was fabricated and examined with tensile test due to its geometrical profile. For the FE computation, FE model with CAX4R element was made according to the dimension size of rod specimen of welded zone as shown in Figure 9.9. In addition, there are 5538 and 5300 nodes and elements for the FE model of rod specimen of welded zone, respectively, and the area of measured region is only 0.1×0.1 mm^2.

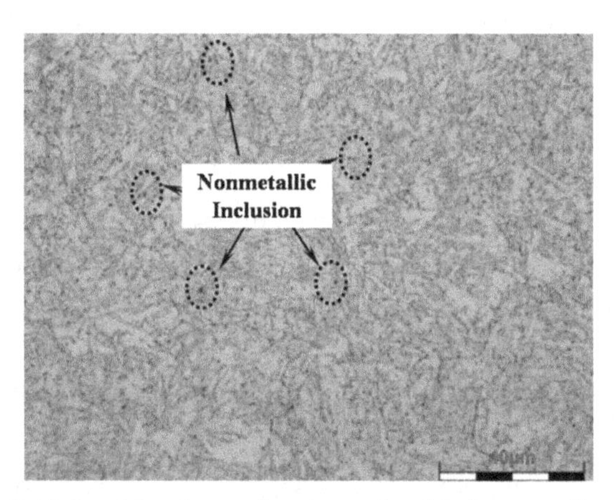

FIGURE 9.8 Initial void volume fraction of welded zone of examined butt-welded joint

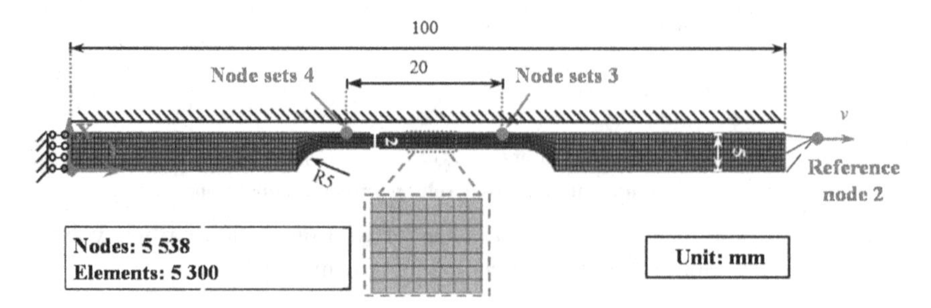

FIGURE 9.9 FE model of rod specimen of welded zone tensile test

Similarly, the nodes in the clamping region of extensometer during rod specimen tensile test were selected to be node sets 3 and 4 as indicated in Figure 9.9, and the FE model of rod specimen of welded zone was constrained to prevent the movement and rotation, and the tensile loading was applied to reference nodes 2. In addition, the axial symmetrical boundary condition was employed. After FE computation, the results of tensile loading of reference nodes 2 as well as relative displacement of node sets 3 and 4 can be obtained, then the relationship between stress and strain of rod specimen of welded zone can be established as shown in Figure 9.10.

With the GTN damage model, elastic modulus of 192 GPa, Poisson's ratio of 0.3, and density of 7860 kg/m^3 were considered. Explicit dynamic analysis was carried out to examine the tensile test process of rod specimen of welded zone while the tensile speed is about 5 mm/min. There are seven cases of FE computation with void nucleation particle volume fraction (f nucleation) of 0.015, 0.02, 0.025, 0.03, 0.035, 0.04, and 0.045. Comparison between computed engineering stress and strain curves of welded zone with different values of void nucleation particle volume fraction (f nucleation) and measured results is demonstrated in Figure 9.8. It can be seen that there is a good agreement between the computed results and measured data, while the void nucleation particle volume fraction (f nucleation) of welded zone is essential. When the engineering strain is less than 0.13, the

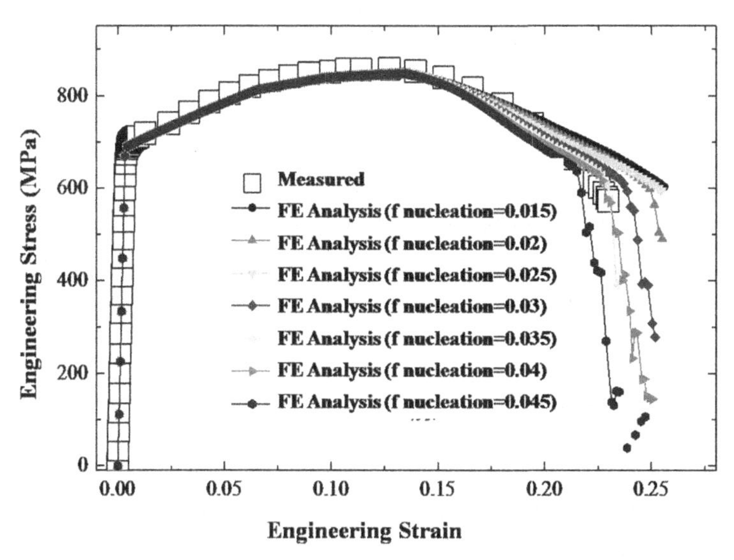

FIGURE 9.10 Comparison of engineering stress and strain curves of welded zone between computed and measured results

almost identical computed results can be observed with different values of void nucleation particle volume fraction (f nucleation).

Similarly, variance as defined by Eq. (9.1) was calculated to closely evaluate the influence of void nucleation particle volume fraction (f nucleation) on engineering stress and strain curve of rod specimen of welded zone. As shown in Figure 9.11, the variance between computed and measured results will decrease and then increase with the increasing of void nucleation particle volume fraction (f nucleation), and the variance will be the lowest when the void nucleation particle volume fraction (f nucleation) of welded zone equals 0.04 (Figure 9.10).

With the above FE computation and variance analysis, parameters of GTN damage model for base material and welded zone of butt-welded joint of Q690 thick plate were summarized in Table 9.3. It can be found that the initial void volume fraction (f0) of welded zone is 40 times of that of base material.

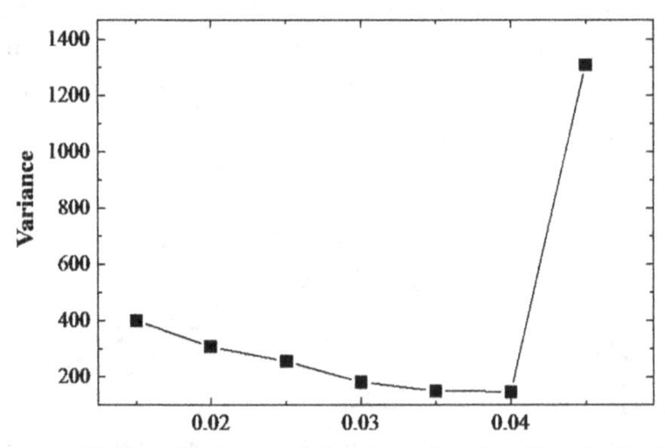

FIGURE 9.11 Variance with different values of void nucleation particle volume fraction (f nucleation) of rod specimen of welded zone

TABLE 9.3 Parameters of GTN damage model for base material and welded zone.

	q_1	q_2	q_3	f_0	f Critical	f Fracture	ε Nucleation	S Nucleation	f Nucleation
Base Material	1.5	1.0	2.25	2.5E-4	1.5E-1	1.7E-1	3.0E-1	1.0E-1	1.5E-2
Welded Zone	1.5	1.0	2.25	1.0E-2	1.5E-1	1.7E-1	3.0E-1	1.0E-1	4.0E-2

9.2.3 GTN Model Validation with Welded Joint Specimen

Based on the dimension of rectangle specimen of butt-welded joint for the tensile test as demonstrated in Figure 9.2, FE model with solid elements of examined welded joint specimen was made as shown in Figure 9.12, while there are 15,150 nodes and 11,950 elements.

In addition, the region with red color is the welded metal, and other region is all base material. Similar to previous investigation, the nodes in the clamping region of extensometer during rectangle specimen tensile test of welded joint were selected to be node sets 5 and 6 as indicated in Figure 9.12, and the FE model of rectangle specimen of welded joint was constrained to prevent the movement and rotation, and the tensile loading was applied to reference nodes 3.

After FE computation, the results of tensile loading of reference nodes 3 as well as relative displacement of node sets 5 and 6 can be obtained, then the relationship between stress and strain of rod specimen of butt-welded joint can be established as shown in Figure 9.13. In detail, applying the parameters value of GTN damage model of base material and welded zone to the corresponding elements, the tensile test speed of reference nodes 3 is about 5 mm/min, and explicit dynamic analysis was carried out to examine the tensile test behavior of rectangle specimen of butt-welded joint while the tensile speed is about 5 mm/min.

During the FE computation, the element will fail due to fracture when its void volume fraction reaches the void volume fraction of fracture occurrence (f fracture). In addition, due to the size of rectangle specimen of butt-welded joint as well as its fabrication process with cutting and polishing, the welding residual stress can be ignored during FE computation of fracture response.

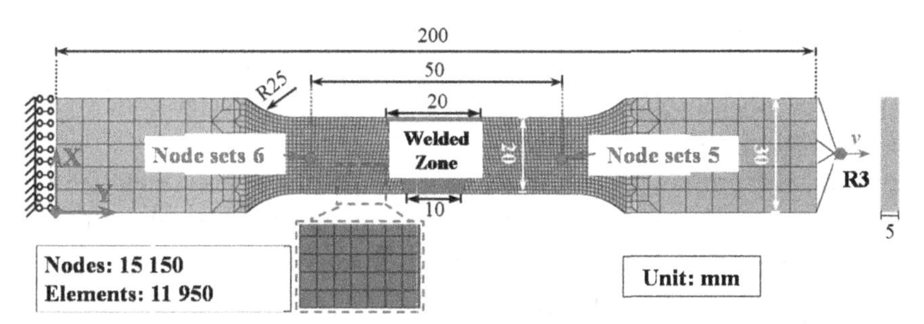

FIGURE 9.12 FE model of rectangle specimen of butt-welded joint tensile test

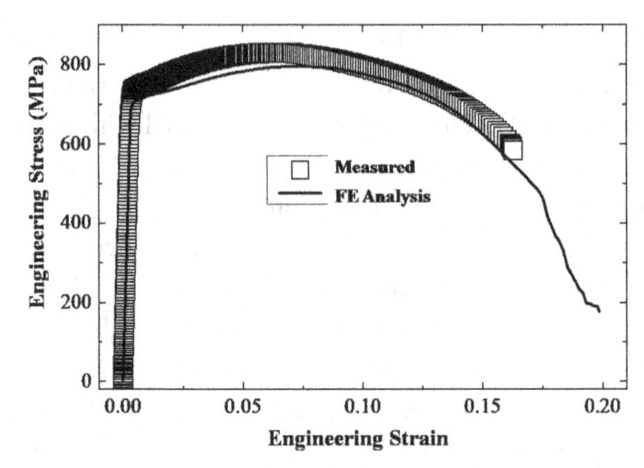

FIGURE 9.13 Comparison of engineering stress and strain curves of welded joint specimen between computed and measured results

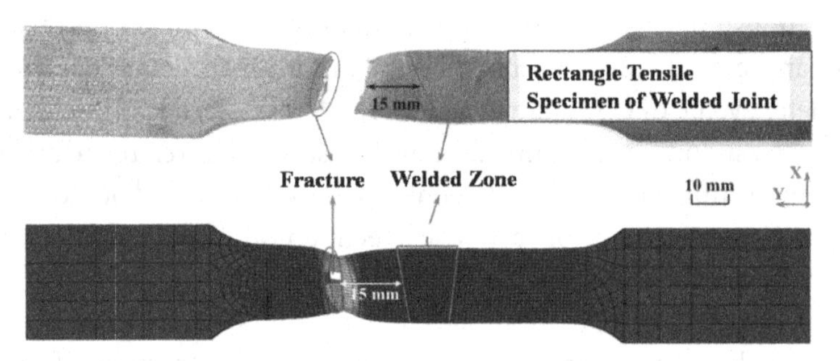

FIGURE 9.14 Comparison of fracture behaviors of axial tensile specimen of butt-welded joint by experiment and computation

As shown in Figure 9.14, the fracture position occurs away from fusion line with average distance of 15 mm, and the computed results were then compared with measurement of tensile test rectangle specimen of examined butt-welded joint. There is good agreement between computed and measured results with the proposed parameters of GTN model as well as engineering stress-strain curves, and the difference may result from the hypothesis that mechanical performance of the heat-affected zone (HAZ) is identical to that of base material.

Thus, the proposed parameters of GTN model of base material and welded zone of butt-welded joint of Q690 thick plate were validated, and after fabrication of tensile test specimen of examined butt-welded joint, residual stress due to welding of tensile test specimen can be neglected. Later, the influence of welding residual stress on fracture performance of butt-welded joint of Q690 thick plate will be examined.

9.3 ASSESSMENT OF FRACTURE PERFORMANCE OF BUTT-WELDED JOINT

Based on the previous investigation, the initial void volume fractions (f0) of base material and welded zone of examined butt-welded joint of Q690 thick plate are different. Based on the previous investigation, the initial void volume fractions (f0) of base material and welded zone of examined butt-welded joint of Q690 thick plate are different, while initial void volume fraction (f0) of base material comes from the initial material defect and initial void volume fraction (f0) of welded zone dominantly results from microflaw due to welding processing.

Welding defect will easily cause strength degradation of welded joint or welded structures compared with base material, which is also the origin of fracture damage. Therefore, there are four computational cases as summarized in Table 9.4 during the FE computation of transverse tensile response of examined butt-welded joint of Q690 thick plate.

With the FE model with solid element of butt-welded joint of Q690 thick plate as mentioned in Figure 8.3 for transient thermal analysis, FE computation for transverse stretching was then carried out with the application of material properties (such as elastic modules, passion ratio and density) and parameters of GTN damage model as summarized in Table 9.2 to the corresponding element of FE model of butt-welded joint of Q690 thick plate.

The contour of initial void volume fraction (f0) of butt-welded joint of Q690 thick plate for abovementioned four computational cases was demonstrated in Figure 9.15, and the computed transverse welding residual

TABLE 9.4 Computational cases of FE computation of transverse tensile response of butt-welded joint of Q690 thick plate.

	Description
Case 1	Initial void volume fraction (f0) of welded zone is identical to that of base material, while the influence of welding defects is ignored, and the influence of transverse welding residual stress is not considered
Case 2	Only influence of welding defects is considered
Case 3	Only influence of transverse welding residual stress is considered, and initial void volume fraction (f0) of welded zone is identical to that of base material while the influence of welding defect is ignored
Case 4	Both influences of welding defects and transverse welding residual stress are considered

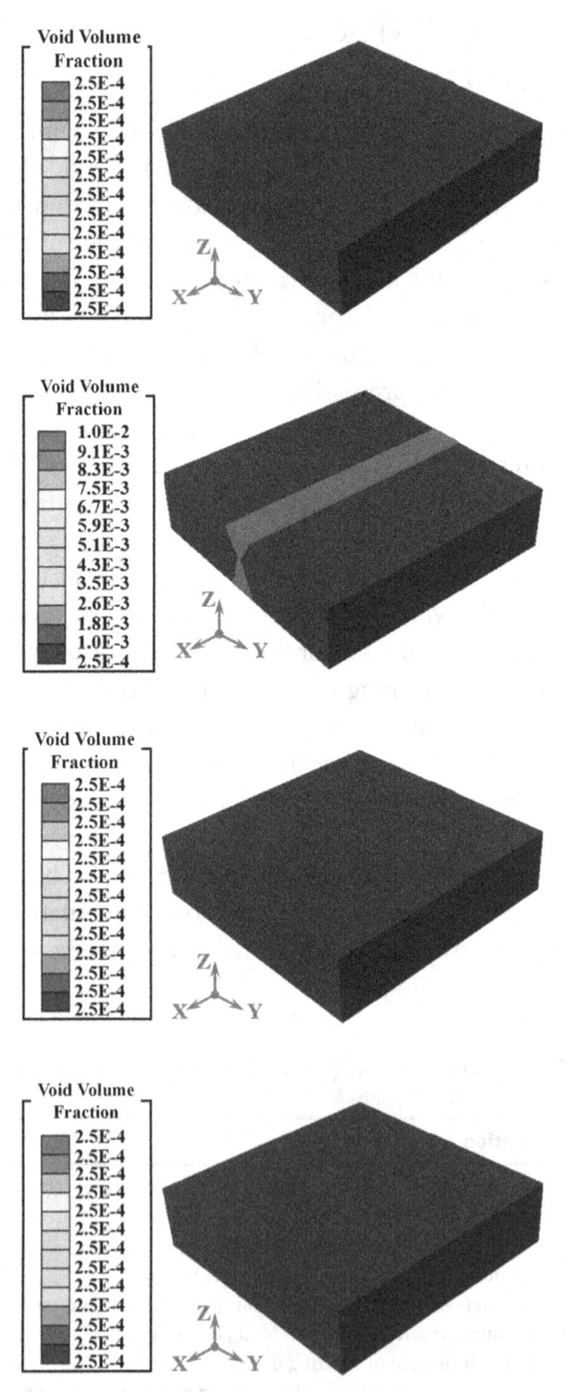

FIGURE 9.15 Initial void volume fraction (f0) of examined butt-welded joint for the different computational cases under transverse tension

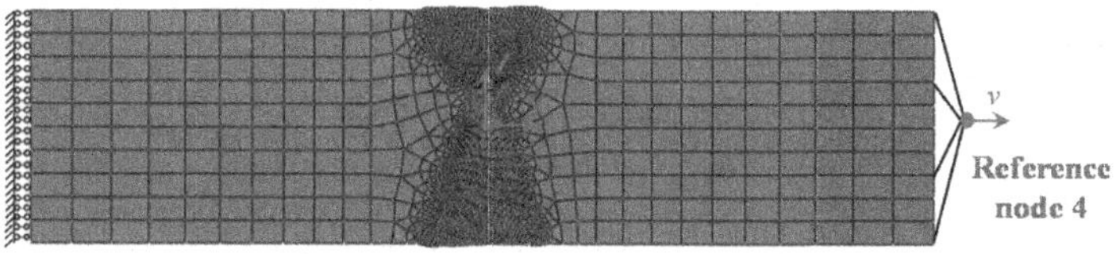

FIGURE 9.16 Boundary condition and external loading for transverse tensile of butt-welded joint of Q690 thick plate

stress as shown in Figure 8.23 was considered to be initial stress during FE computation of transverse tensile, which will be employed for the Case 3 and Case 4. In addition, during the FE computation of transverse tensile of butt-welded joint of Q690 thick plate, boundary condition was fixed to prevent the movement and rotation of FE model, and the external loading was coupled to the reference node 4 as indicated in Figure 9.16.

As shown in Figure 9.17, evolutions of transverse stress of butt-welded joint of Q690 thick plate during transverse stretching were demonstrated for computational case 2 considering welding defect and case 4 considering both transverse residual stress and welding defects.

In addition, Figure 9.17(a) shows that there is no transverse stress in the butt-welded joint of Q690 thick plate at the beginning of transverse tensile and its magnitude is zero. With the increasing transverse tensile loading, the transverse stress of butt-welded joint of Q690 thick plate will increase. When the transverse tensile loading reaches the yield strength of Q690 high tensile strength steel, the region except for the top and bottom surfaces of welded zone of butt-welded joint of Q690 thick plate will yield plastic strain hardening.

During the transverse tensile, stress concentration will be generated at the region of top and bottom surfaces of welded zone due to nonuniform geometrical features. Meanwhile, with Figure 9.17(b), there is much-complicated distribution of transverse stress in the examined butt welded joint of Q690 thick plate at the beginning of transverse tensile. With the increasing of transverse tensile loading, transverse stress in the middle region of base material of butt-welded joint of Q690 thick plate along the thickness direction will firstly increase to the magnitude of yield stress.

Moreover, transverse stress at the top and bottom surfaces of region of base material as well as middle region of welded zone will be converted from compressive stress to tensile stress under the external loading. It can be seen that transverse residual stress will significantly influence the

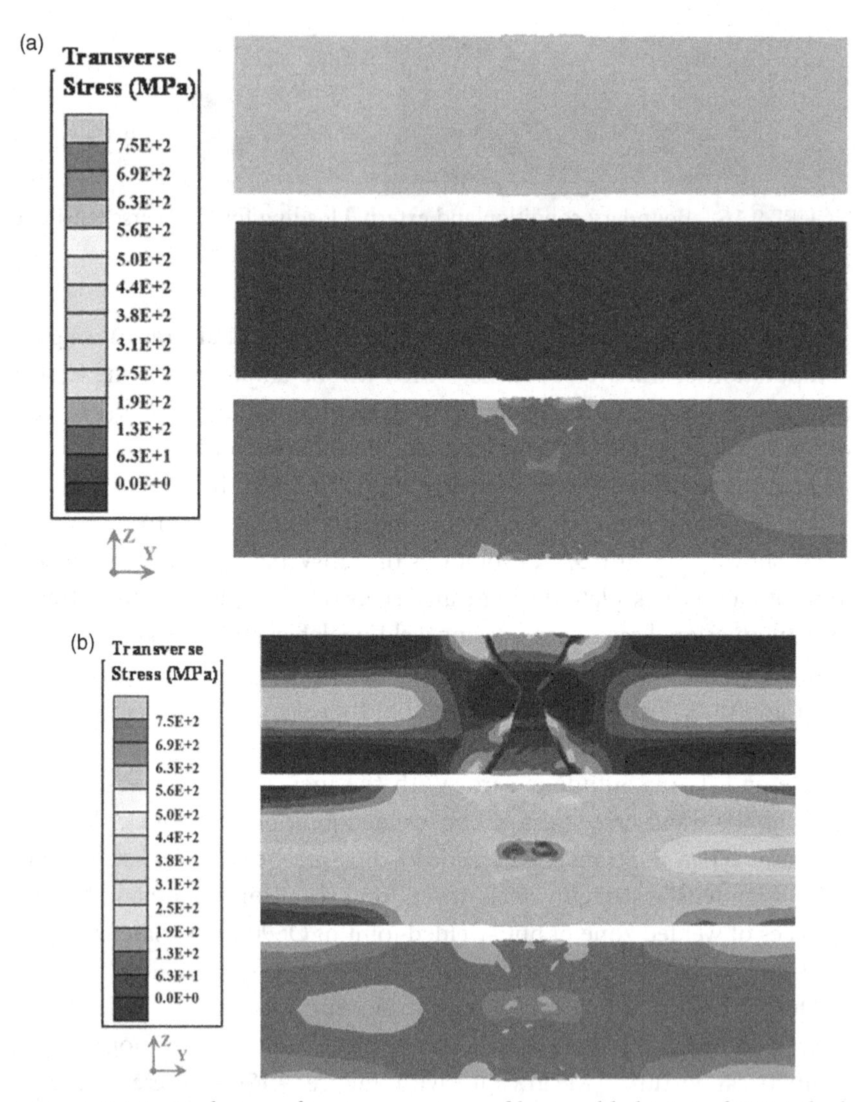

FIGURE 9.17 Evolution of transverse stress of butt-welded joint of Q690 thick plate during transverse stretching

distribution of transverse stress of butt-welded joint of Q690 thick plate during elastic response stage under transverse stretching. Middle region of base material will yield beforehand due to the superposition of welding tensile residual stress and external loading, while compressive residual stress will influence the effect of external loading to delay yielding of local

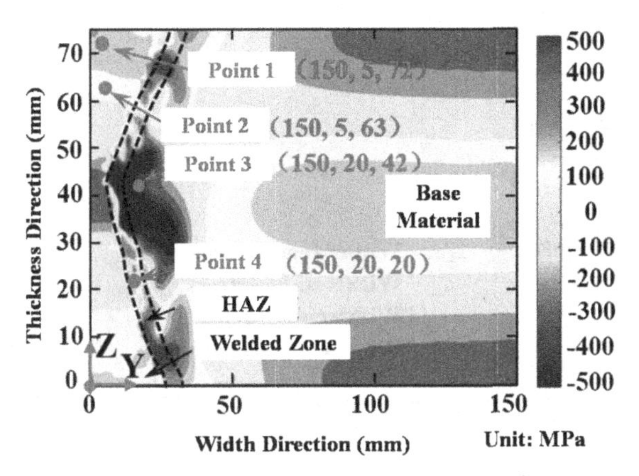

FIGURE 9.18 Position of examined points of fracture performance

region with welding compressive residual stress of butt-welded joint of Q690 thick plate.

As shown in Figure 9.18, there are four points to be considered for their mechanical responses: while points 1 and 2 are located in the region of welded zone, point 3 is located in the region of base material near heat-affected zone (HAZ), and point 4 is located in the region of HAZ near the fusion line.

Moreover, the true stress and strain curves of examined points on the middle cross-section of butt-welded joint of Q690 thick plate as indicated in Figure 9.18 are demonstrated in Figure 9.19. It can be seen that the relations of true stress and strain of examined points of butt-welded joint of Q690 thick plate were influenced by both welding transverse residual stress and welding defect. In addition, welding defect will influence the maximal value of true stress and degrade the fracture strength of butt-welded joint of Q690 thick plate. Welding transverse residual stress with compressive characteristics will improve the maximal value of true stress as well as the fracture strength.

It is difficult to evaluate the fracture performance of butt-welded joint of Q690 thick plate with the relation of true stress and strain during transverse tensile process, and relations of engineering stress and strain of examined points as indicated in Figure 9.19 were obtained and are demonstrated in Figure 9.20. Without considering the influences of welding transverse residual stress and welding defects, the magnitudes of ultimate and fracture strength of butt-welded joint of Q690 thick plate are 825 MPa and 716 MPa, respectively. It can be seen that the fracture performance of

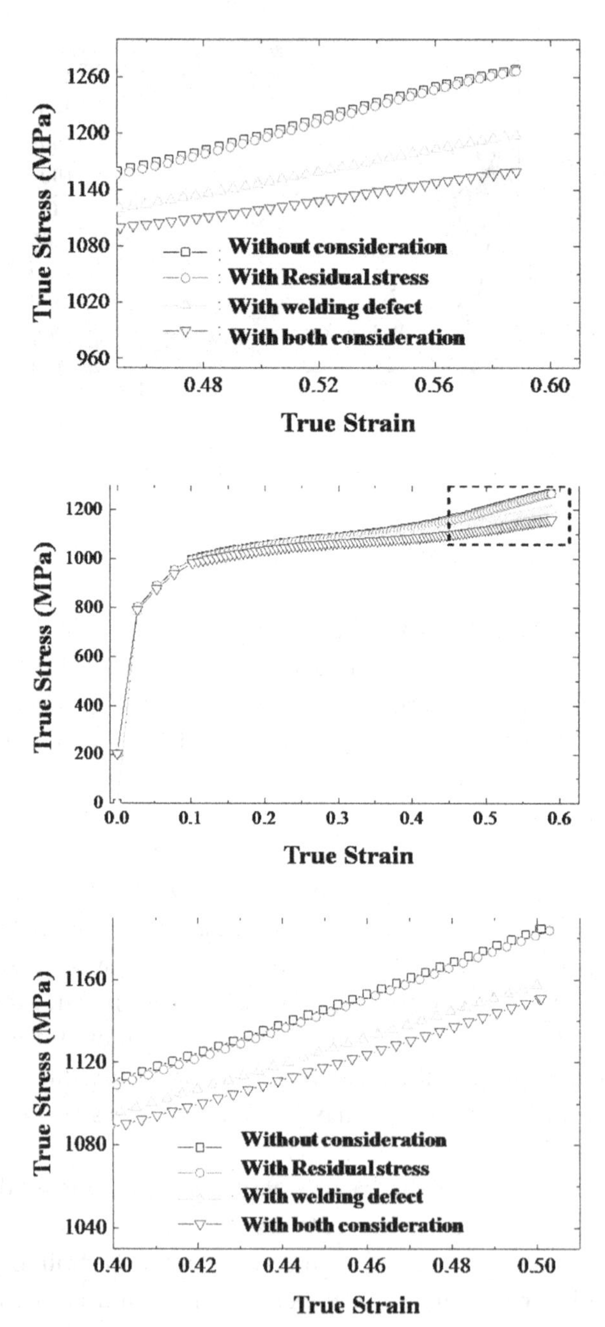

FIGURE 9.19 Relation of true stress and strain of examined points of butt-welded joint of Q690 thick plate

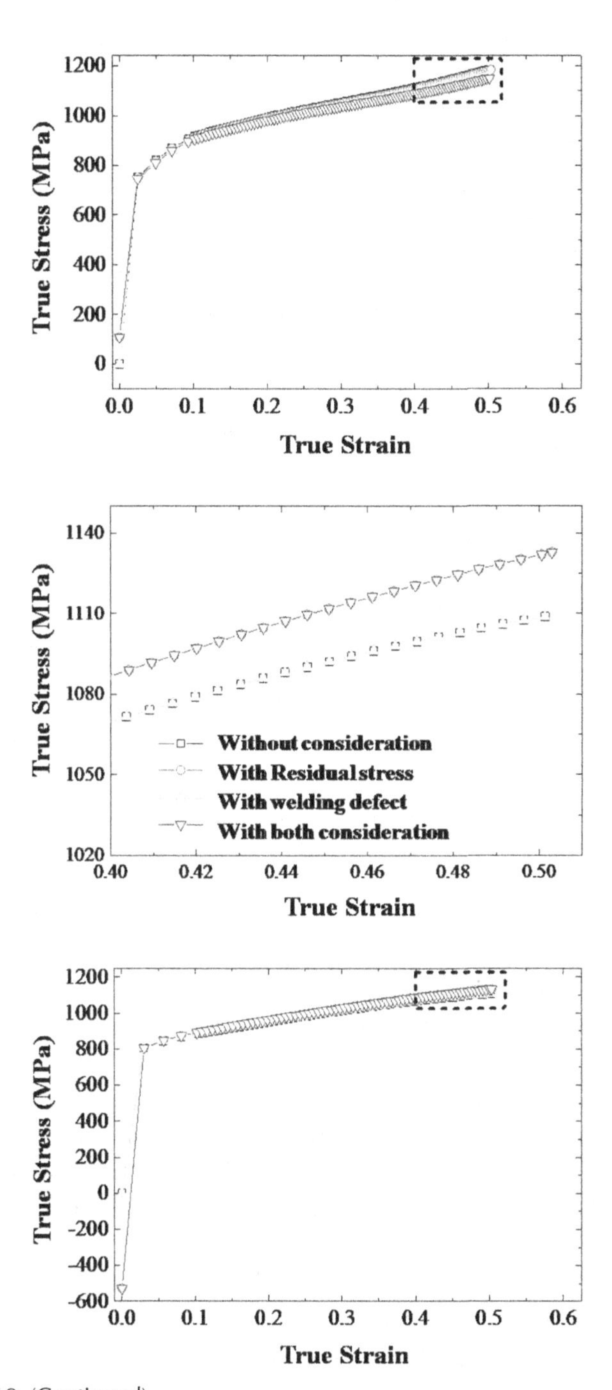

FIGURE 9.19 (Continued)

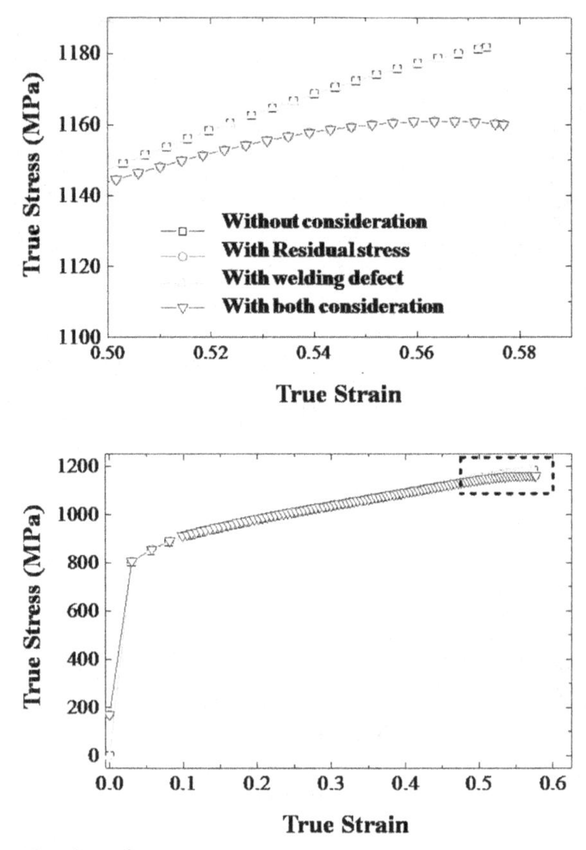

FIGURE 9.19 (Continued)

butt-welded joint of Q690 thick plate was influenced by welding transverse residual stress and welding defects together.

In detail, the influence of welding transverse residual stress with magnitude of tensile stress of 200 MPa on fracture performance of butt-welded joint of Q690 thick plate can be ignored when the welding defect was only considered. Initial void volume fraction (f0) of welded zone will be 0.01 from the value of 0.00025 for base material, and the magnitudes of ultimate and fracture strength will decrease to 886 MPa and 665 MPa, respectively.

Moreover, fracture strength of butt-welded joint of Q690 thick plate will decrease from 704 MPa to 640 MPa by about 9% when the influences of welding transverse residual stress and welding defects are both considered. Fracture strength of butt-welded joint of Q690 thick plate will increase from 670 MPa to 685 MPa by about 2% when the welding transverse residual stress is about −520 MPa.

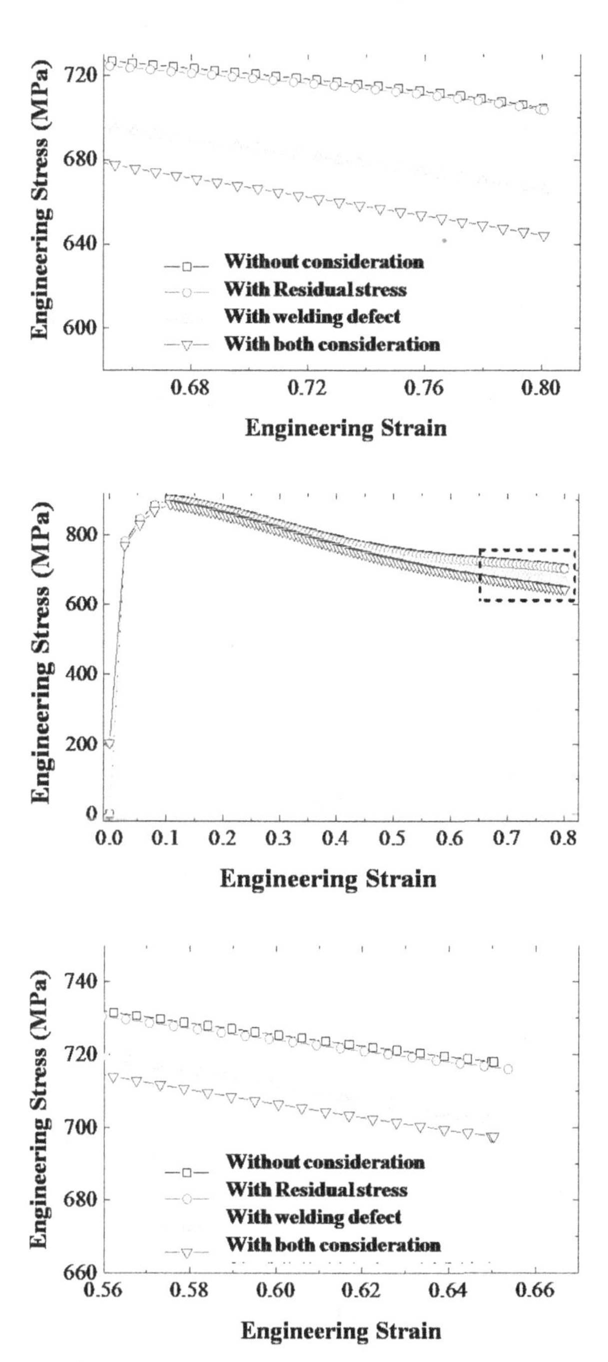

FIGURE 9.20 Relation of engineering stress and strain of examined points of butt-welded joint of Q690 thick plate

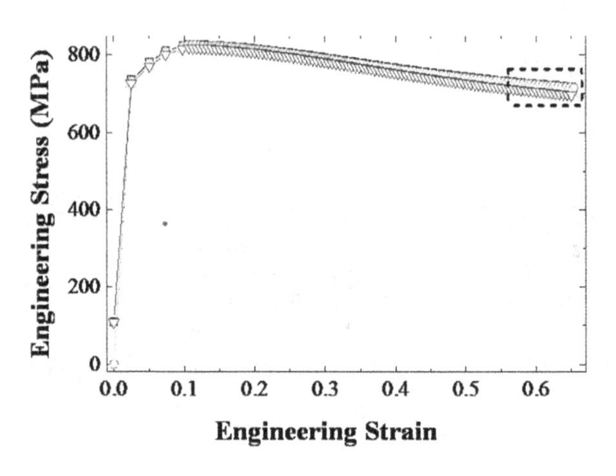

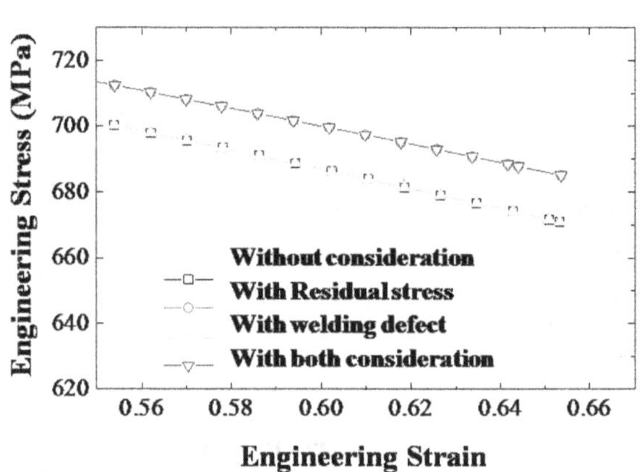

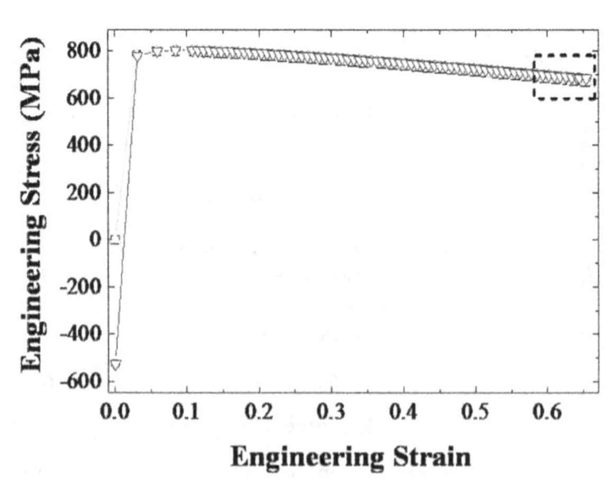

FIGURE 9.20 (Continued)

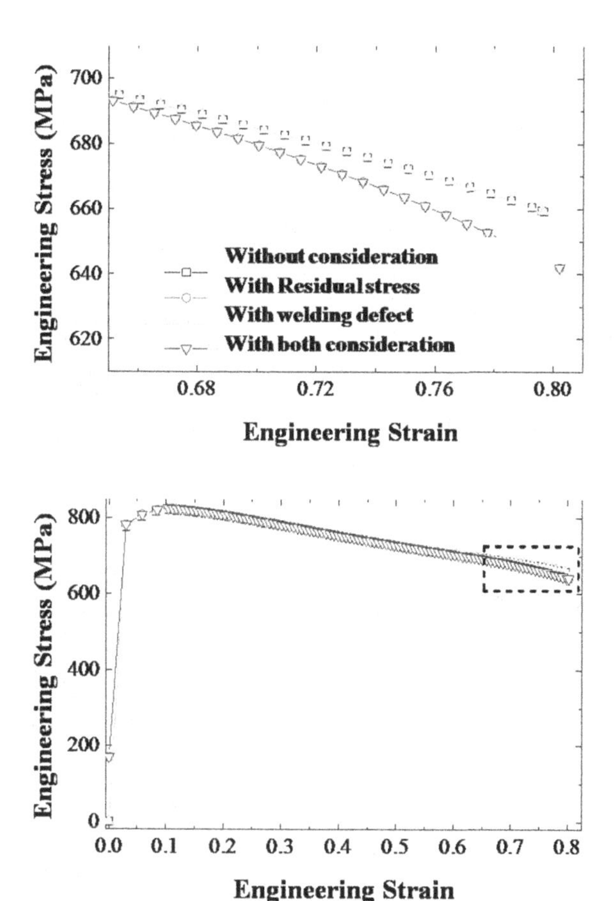

FIGURE 9.20 (Continued)

9.4 MECHANISM ANALYSIS WITH GTN MODEL

With the previous investigation, influence of welding transverse residual stress and welding defects on stress-strain curve of butt-welded joint of Q690 thick plate were examined and demonstrated. In order to clarify and understand the generation mechanism, the influence of welding transverse residual stress and welding defects on the void volume fraction of fracture occurrence (f fracture) as well as void evolution process will be considered to examine the influence mechanism of welding transverse residual stress on fracture performance.

During the FE computation of butt-welded joint of Q690 thick plate under transverse stretching loading, element will totally lose carrying capacity and there will be position of fracture occurrence when void volume

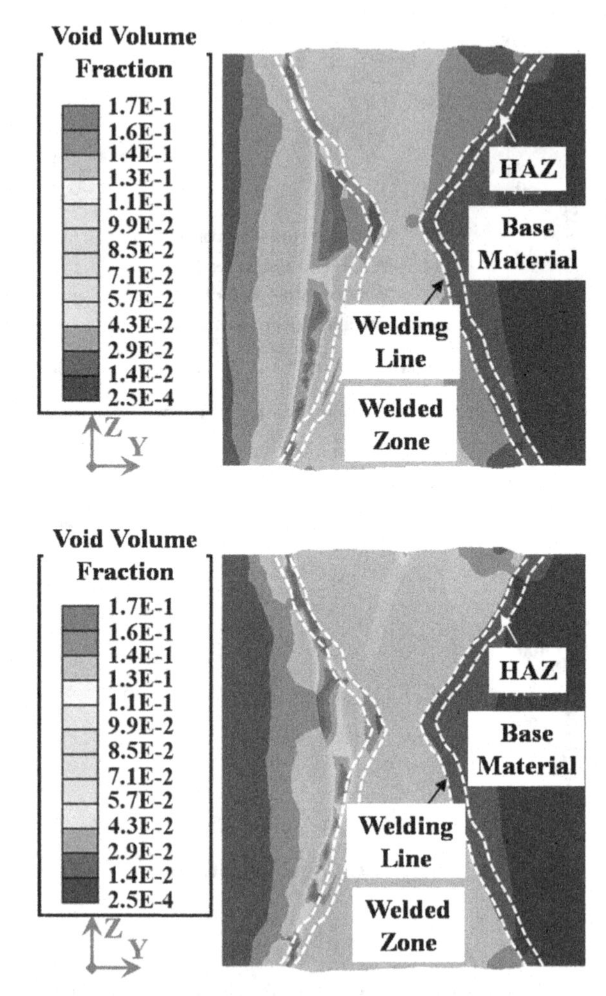

FIGURE 9.21 Contour of void volume fraction of fracture occurrence (f fracture) of butt-welded joint of Q690 thick plate during transverse stretching

fraction of this element reaches the magnitude of the void volume fraction of fracture occurrence (f fracture). Due to the complex mechanical behavior of initialization and propagation of fracture and loss of influential factors, distribution of the void volume fraction of fracture occurrence (f fracture) will be employed to directly examine the influence of welding transverse residual stress and welding defects on fracture performance of butt-welded joint of Q690 thick plate. As shown in Figure 9.21, the contour of the void volume fraction of fracture occurrence (f fracture) of butt-welded joint of

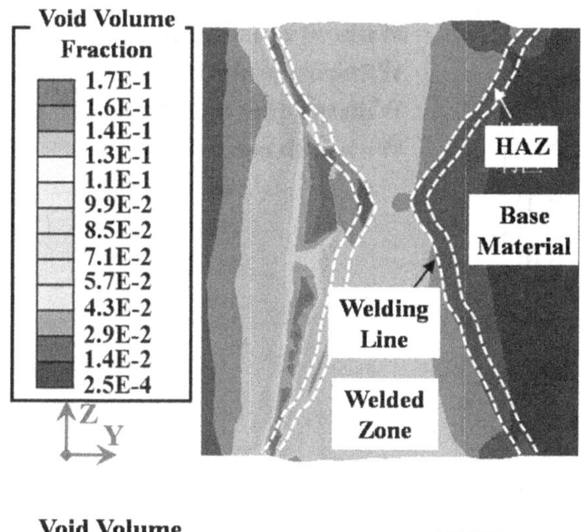

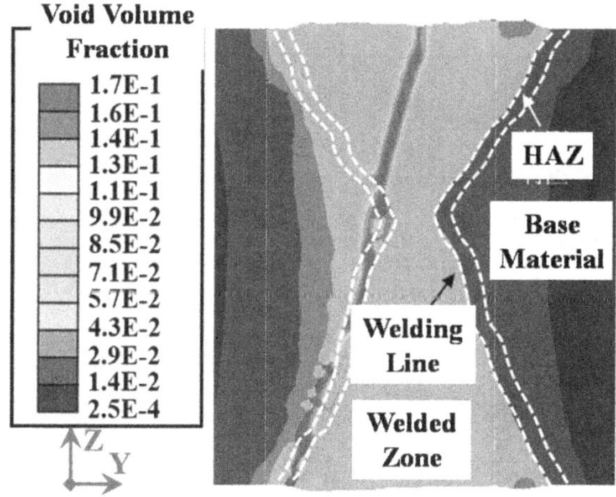

FIGURE 9.21 (Continued)

Q690 thick plate under transverse stretching with four computational cases as summarized in Table 9.4 was demonstrated.

When the influence of only welding transverse residual stress or welding defects was considered, it can be seen that the region with large void volume fraction of fracture occurrence (f fracture) is dominantly located at the region of base material of butt-welded joint of Q690 thick plate. However, the region with large void volume fraction of fracture occurrence (f fracture) dominantly is located at the region of welded zone and heat-affected zone (HAZ) of butt-welded joint of Q690 thick plate, when the

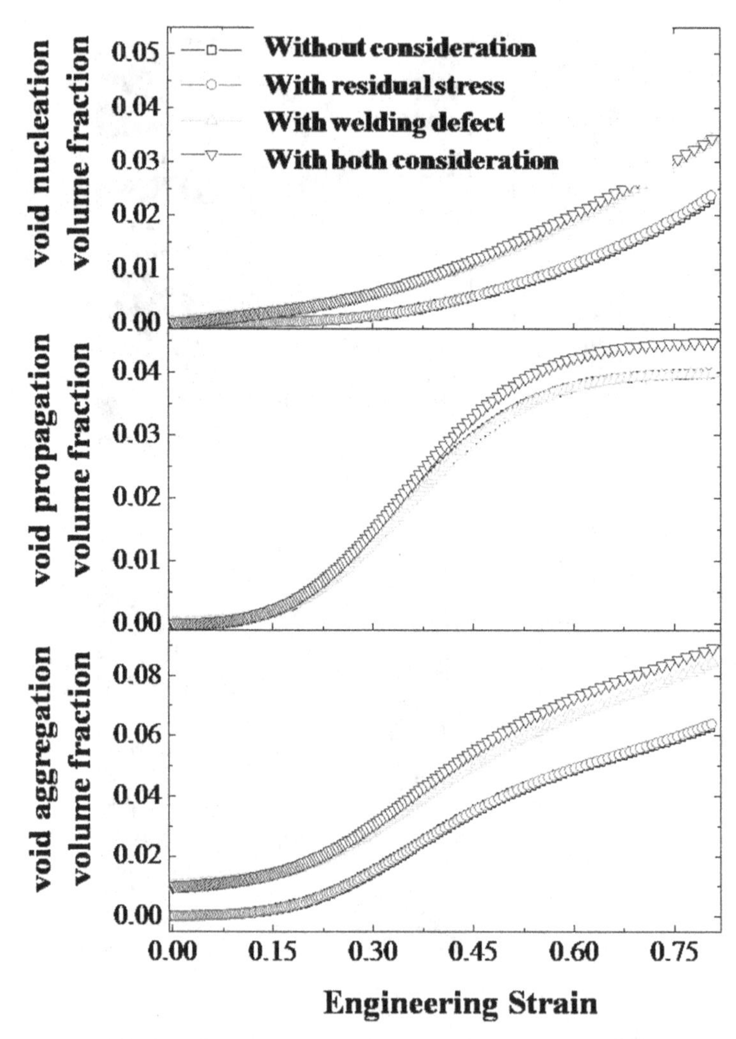

FIGURE 9.22 Void evolution process of examined points on middle cross-section of butt-welded joint of Q690 thick plate

influences of welding transverse residual stress and welding defects were both considered.

With the GTN damage model, the void evolution process of initialization, propagation, and aggregation of metal material is the elementary cause of eventually generating fracture failure of examined metal material under the external loading. In general, the initial void volume fraction (f0) of heat-affected zone (HAZ) is identical to that of base material, which will be much less than that of welded zone.

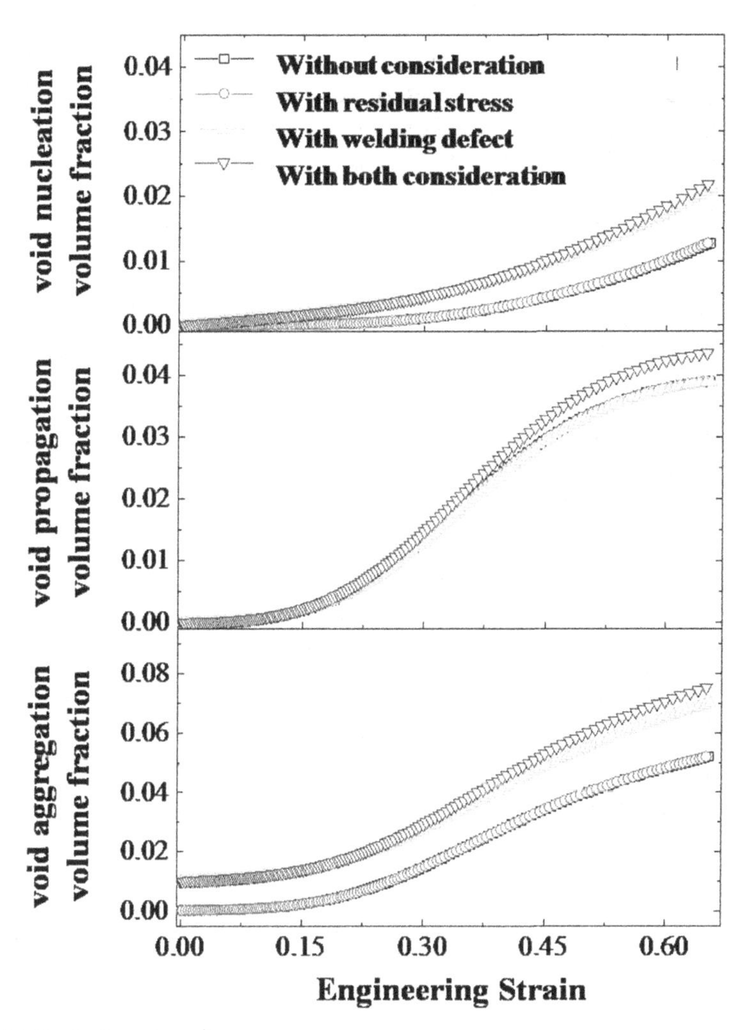

FIGURE 9.22 (Continued)

Taking the examined points as indicated in Figure 9.18 as research object, the influence of welding transverse residual stress on void evolution process was examined, and the influence mechanism of welding transverse residual stress on fracture performance of butt-welded joint of Q690 thick plate will be then clarified. As shown in Figure 9.22, welding transverse tensile residual stress has almost no influence on void initialization during the transverse tensile process of butt-welded joint of Q690 thick plate. When the magnitude of transverse tensile residual stress is 200 MPa and the influence of welding defects was considered, volume fraction of

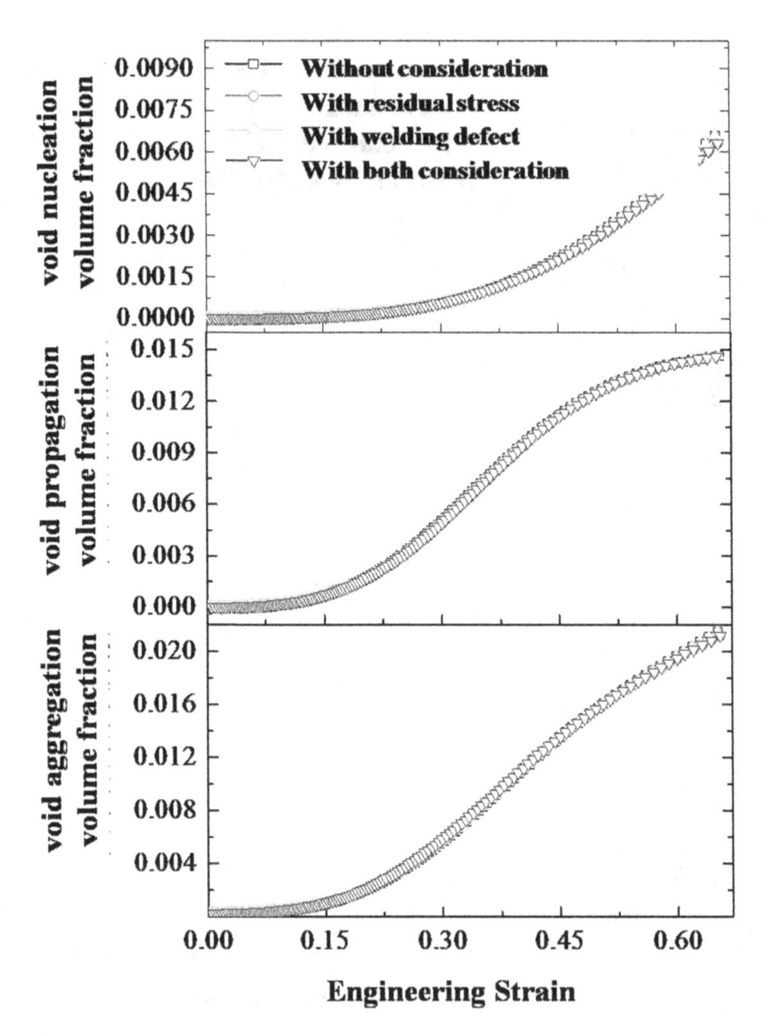

FIGURE 9.22 (Continued)

void propagation increased from 0.03989 to 0.04481 by about 12.3%, and volume fraction of void aggregation increased from 0.06318 to 0.08914 by about 41.1%. However, when the magnitude of transverse tensile residual stress is 170 MPa without the influence of welding defects, there will be no variation in void evolution process of initialization, propagation, and aggregation compared with that without considering the transverse residual stress and welding defect. When the welding transverse residual stress is compressive stress, volume fraction of void initialization decreased

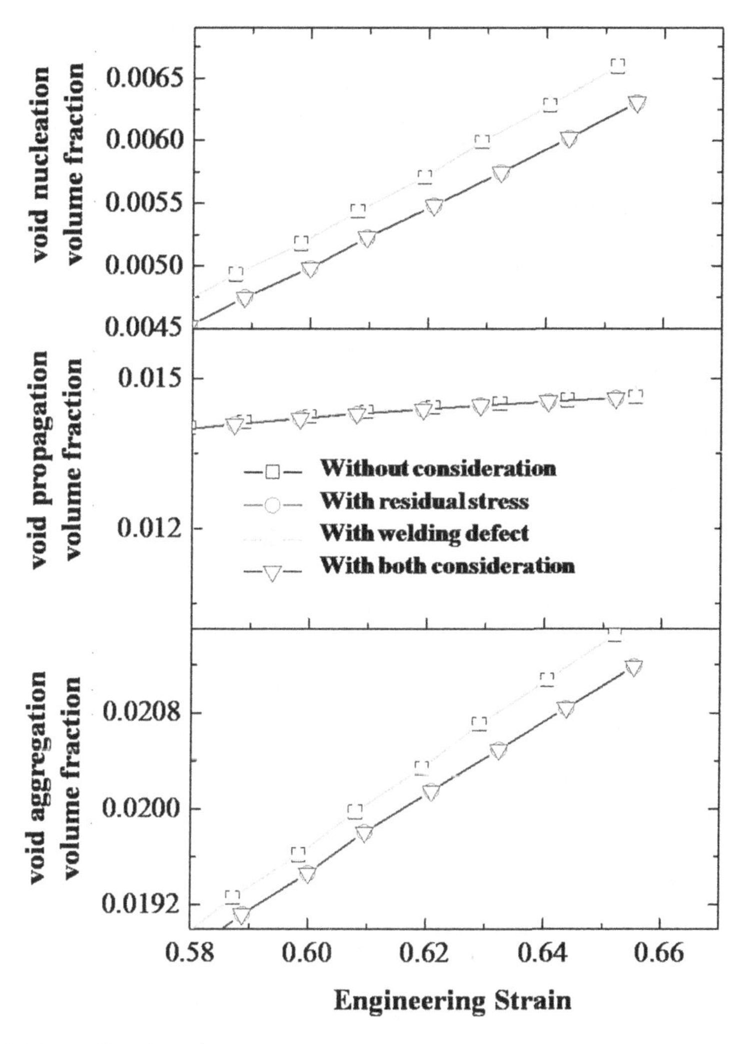

FIGURE 9.22 (Continued)

from 0.0066 to 0.0063 with 5%, and volume fraction of void aggregation also decreased from 0.02145 to 0.02119 with 1.2%.

With the above investigation, during the transverse tensile process of butt-welded joint of Q690 thick plate welding, transverse residual stress in the region of welded zone will result in significant degradation of its fracture performance by means of increasing the process of void propagation and aggregation, as well as rapid increasing of void volume fraction to the void volume fraction of fracture occurrence (f fracture). However, welding transverse compressive residual stress will restrain the void evolution

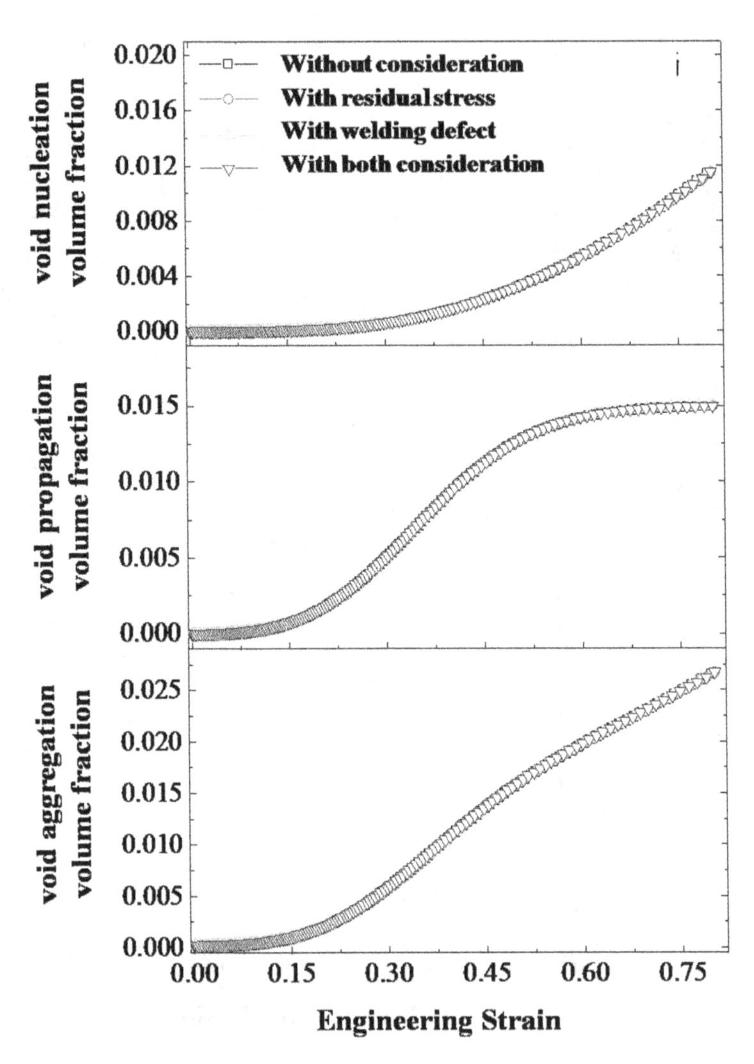

FIGURE 9.22 (Continued)

process of initialization, propagation, and aggregation and result in a slowly increasing void volume fraction, which will be beneficial to enhance the fracture performance of butt-welded joint of Q690 thick plate.

In conclusion, it is important to concentrate on the distribution of welding residual stress of welded joint or welded structure after the multipass welding of thick plate with high tensile strength steel. Relaxation practice of welding residual stress will be employed to decrease magnitude of tensile residual stress in the region of welded zone and heat-affected zone

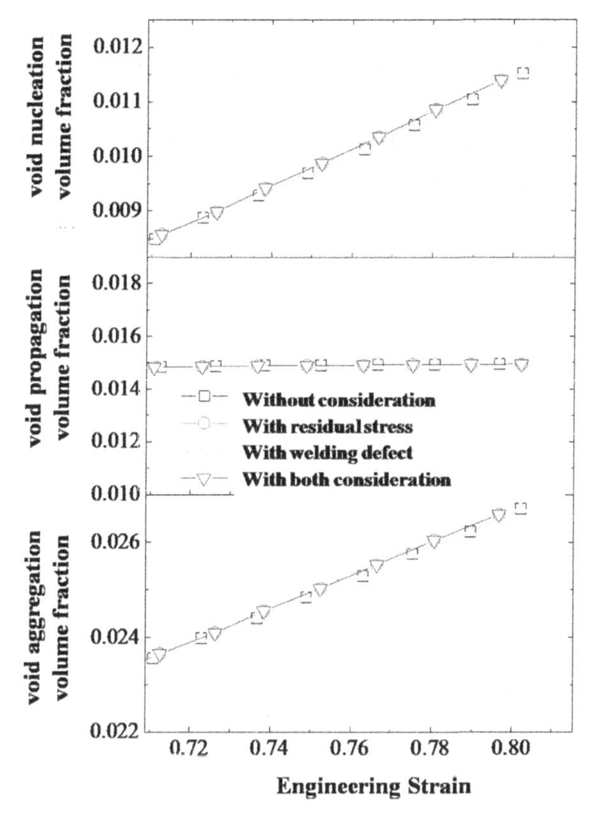

FIGURE 9.22 (Continued)

(HAZ); moreover, that tensile residual stress is converted to be compressive residual stress in the local region of welded zone and heat-affected zone (HAZ) will be beneficial to improve the fracture performance of welded joint and welded structure with thick plate of high tensile strength steel, while the region with stress concentration will be avoided to be origin of fracture.

9.5 SUMMARY

With the axial tensile test experiment of specimen, parameters of GTN damage model were confirmed by means of FE computation, then fracture performance of examined butt-welded joint specimen of Q690 thick plate was evaluated with GTN damage model, which has a good agreement compared with experiment. Fracture performance of examined butt-welded joint was numerically examined with four computational cases, while the influences of transverse residual stress and welding defects are individually considered. Eventually, mechanism of welding transverse residual stress to degrade the fracture performance of examined butt-welded joint was presented with GTN damage model as well as the evolution of void nucleation, propagation, and aggregation.

References

[1] K. Satoh, Y. Ueda and J. Fujimoto (1979). Welding distortion and residual stresses. *Sanpo Publications*, Tokyo, Japan.

[2] D. Radaj (1992). Heat effects of welding: temperature field, residual stress and distortion. *Springer Verlag Publishing*, Berlin, Heidelberg, Germany.

[3] Y. Ueda and T. Yamakawa (1971). Analysis of thermal elastic plastic stress and strain during welding by finite element method. *Transactions of Japan Welding Society*, 2(2), 90–100.

[4] H.D. Hibbitt and P.V. Marcal (1973). A numerical thermo-mechanical model for the welding and subsequent loading of a fabricated structure. *Computers and Structures*, 3(5), 1145–1174.

[5] E. Friedman (1975). Thermal mechanical analysis of the welding process using the finite element method. *Journal of Pressure Vessel Technology*, 97(3), 206–213.

[6] L.E. Lindgren and L. Karlsson (1988). Deformations and stresses in welding of shell structures. *International Journal for Numerical Methods in Engineering*, 25(2), 635–655.

[7] K. Masubuchi (1980). Analysis of welded structures: residual stresses, distortion and their consequences. *Pergamon Press*, Oxford, England.

[8] J.A. Goldak, A. Chakravarti and M. Bibby (1984). A new finite element model for welding heat sources. *Metallurgical Transactions B*, 15(2), 299–305.

[9] Y. Ueda and H. Murakawa (1985). Applications of computer and numerical analysis techniques in welding research. *Materials and Design*, 6(3), 103–111.

[10] L.E. Lindgren (2001). Finite element modeling and simulation of welding (part 1): increased complexity. *Journal of Thermal Stresses*, 24(2), 141–192.

[11] L.E. Lindgren (2001). Finite element modeling and simulation of welding (part 2): improved material modeling. *Journal of Thermal Stresses*, 24(3), 195–231.

[12] L.E. Lindgren (2001). Finite element modeling and simulation of welding (part 3): efficiency and integration. *Journal of Thermal Stresses*, 24(4), 305–334.

[13] L.E. Lindgren (2006). Numerical modeling of welding. *Computer Methods in Applied Mechanics and Engineering*, 195(48–49), 6710–6736.

[14] J.A. Goldak and M. Akhlaghi (2005). Computational welding mechanics. *Springer Science & Business Media* , New York, USA.

[15] Y. Ueda (1999). Computational welding mechanics (selected from papers published in English). *Joining and Welding Research Institute, Osaka University*, Osaka, Japan.

[16] L.E. Lindgren (2007). Computational welding mechanics: thermomechanical and microstructural simulations. *Woodhead Publishing Limited*, Cambridge, England.

[17] P. Michaleris and A. Debiccari (1997). Prediction of welding distortion. *Welding Journal*, 76(4), 172–181.

[18] A. Bachorski, M.J. Painter A.J. Smailes and M.A. Wahab (1999). Finite element prediction of distortion during gas metal arc welding using the shrinkage volume approach. *Journal of Materials Processing Technology*, 92–93, 405–409.

[19] G.H. Jung and C.L. Tsai (2004). Plasticity-based distortion analysis for fillet welded thin-plate T-joints. *Welding Journal*, 83(6), 177–187.

[20] D. Camilleri, T. Comlekci and T.G.F. Gray (2005). Computational prediction of out-of-plane welding distortion and experimental investigation. *Journal of Strain Analysis*, 40(2), 161–176.

[21] W.T. Cheng (2005). In-plane Shrinkage strain and their effects on welding distortion in thin-wall structure. *Doctoral Dissertation*, Ohio State University.

[22] L. Zhang, P. Michaleris and P. Marugabandhu (2007). Evaluation of applied plastic strain methods for welding distortion prediction. *Journal of Manufacturing Science and Engineering*, 129(6), 1000–1010.

[23] G.H. Jung (2007). A shell-element-based elastic analysis predicting welding induced distortion for ship panels. *Journal of Ship Research*, 51(2), 128–136.

[24] Y.P. Yang, H. Castner and N. Kapustka (2011). Development of distortion modeling methods for large welded structures. *Journal of Ship Production and Design*, 27(1), 26–34.

[25] H. Nishikawa, H. Serizawa and H. Murakawa (2007). Actual application of FEM to analysis of large scale mechanical problems in welding. *Science and Technology of Welding and Joining*, 12(2), 147–152.

[26] H. Nishikawa (2006). Development of FEM for analysis of large scale mechanical problems in welding and its application to practical problems. *Doctoral Dissertation*, Osaka University.

[27] Murakawa H, Ma N, Huang H (2015). Iterative substructure method employing concept of inherent strain for large-scale welding problems. *Welding in the World*, 59, 53–63.

[28] K. Ikushima, M. Shibahara, S. Itoh and N. Ma (2012). Ultra large scale FE computation using idealized explicit FEM accelerated by GPU. *Transactions of JWRI*, Special Issue on WSE2011, 91–92.

[29] K. Ikushima, S. Itoh and M. Shibahara (2015). Development of idealized explicit FEM using GPU parallelization and its application to large-scale analysis of residual stress of multi-pass welded pipe joint. *Welding in the World*, 59, 589–595.

[30] N. Ma (2016). An accelerated explicit method with GPU parallel computing for thermal stress and welding deformation of large structure models. *International Journal Advanced Manufacturing Technology*, 87(5), 2195–2211.

[31] H. Runnemalm and S. Hyun (2000). Three dimensional welding analysis using an adaptive mesh scheme. *Computer Methods in Applied Mechanics and Engineering*, 189(2), 515–523.

[32] J. Kim, S. Im and H.G. Kim (2005). Finite element analysis of arc welding process by parallel computation. *Transactions of Japan Welding Society*, 23(2), 203–208.

[33] Y. Tian, C. Wang, D. Zhu and Y. Zhou (2008). Finite element modeling of electron beam welding of a large complex Al alloy structure by parallel computations. *Journal of Materials Processing Technology*, 199, 41–48.

[34] Y. Luo, H. Murakawa and Y. Ueda (1997). Prediction of welding deformation and residual stress by elastic FEM based on inherent strain (first report): mechanism of inherent strain production. *Transactions of JWRI*, 26(2), 49–57.

[35] H. Murakawa, Y. Luo and Y. Ueda (1998). Inherent Strain as an interface between computational welding mechanics and its industrial application. *Mathematical Modeling of Weld Phenomena*, 4, 597–619.

[36] Y. Luo, M. Ishiyama and H. Murakawa (1999). Welding deformation of plates with longitudinal curvature. *Transactions of JWRI*, 28(2), 57–65.

[37] Y. Luo, D. Deng, L. Xie and H. Murakawa (2004). Prediction of deformation for large welded structures based on inherent strain. *Transactions of JWRI*, 33(1), 65–70.

[38] D. Deng, H. Murakawa and W. Liang (2007). Numerical simulation of welding distortion in large structures. *Computer Methods in Applied Mechanics Engineering*, 196(45–48), 4613–4627.

[39] R. Wang, J.X. Zhang, H. Serizawa and H. Murakawa (2009). Study of welding inherent deformation s in thin plates based on finite element analysis using interactive substructure method. *Materials and Design*, 30(9), 3474–3481.

[40] J.C. Wang, X. Ma, H. Murakawa, B.G. Teng and S.J. Yuan (2011). Prediction and measurement of welding distortion of a spherical structure assembled form multi thin plates. *Materials and Design*, 32(10), 4728–4737.

[41] K. Masubuchi (1953). Buckling type deformation of thin plate due to welding. *Proceedings of the 3rd Japan National Congress for Applied Mechanics*, Tokyo, Japan, 107–111.

[42] M. Watanabe and K. Satoh (1957). On the type of distortion in various welded joints: shrinkage distortion in welded joint (Report 5). *Journal of Japan Welding Society*, 26(6), 399–405.

[43] M. Watanabe and K. Satoh (1958). Fundamental study on buckling of thin steel plate due to bead-welding. *Journal of Japan Welding Society*, 27(6), 313–320.

[44] M.P. Frank (1975). Buckling distortion of thin aluminum plates during welding. *Master Thesis*, Massachusetts Institute of Technology.

[45] T. Terasaki, K. Maeda, H. Murakawa and T. Nomoto (1998). Critical conditions of plate buckling generated by welding. *Transactions of the Japan Society of Mechanical Engineers (A)*, 64(625), 2239–2244.

[46] N. Toshiharu, T. Terasaki and K. Maeda (1997). Study of parameters controlling weld buckling. *Transactions of the Japan Society of Mechanical Engineers (A)*, 63(609), 1063–1068.

[47] C.L. Tsai, S.C. Park and W.T. Cheng (1999). Welding distortion of a thin-plate panel structure: the effect of welding sequence on panel distortion is evaluated. *Welding Journal*, 78(5), 156–165.

[48] C.L. Tsai, M.S. Han and G.H. Jung (2006). Investigating the bifurcation phenomenon in plate welding. *Welding Journal*, 85(7), 151–162.

[49] P. Michaleris and A. Debiccari (1996). A predictive technique for buckling analysis of thin section panels due to welding. *Journal of Ship Production*, 12(4), 269–275.

[50] P. Michaleris (2011). Minimization of welding distortion and buckling: modelling and implementation. *Woodhead Publishing Limited*, Cambridge, England.

[51] O.A. Vanli and P. Michaleris (2001). Distortion analysis of welded stiffeners. *Journal of Ship Production*, 17(4), 226–240.

[52] M.V. Deo, P. Michaleris and J. Sun (2003). Prediction of buckling distortion of welded structures. *Science and Technology of Welding and Joining*, 8(1), 55–61.

[53] P. Michaleris, L. Zhang, S.R. Bhide and P. Marugabandhu (2007). Evaluation of 2D, 3D and applied plastic strain methods for predicting buckling welding distortion and residual stress. *Science and Technology of Welding and Joining*, 11(6), 707–716.

[54] S.R. Bhide, P. Michaleris, M. Posada and J. Deloach (2006). Comparison of buckling distortion propensity for SAW, GMAW and FSW. *Welding Journal*, 85(9), 189–195.

[55] P. Mollicone, D. Camilleri and T. Gray (2008). Procedural influences on non-linear distortions in welded thin-plate fabrication. *Thin-walled Structures*, 46, 1021–1034.

[56] T.D. Huang, P. Dong, L. Decan, D. Harwig and R. Kumar (2004). Fabrication and engineering technology for lightweight ship structures, part 1: distortions and residual stresses in panel fabrication. *Journal of Ship Production*, 20(1), 43–59.

[57] T.D. Huang, C. Conrardy, P. Dong, P. Keene, L. Kvidahl and L. Decan (2007). Engineering and production technology for lightweight ship structures, part 2: distortion mitigation technique and implements. *Journal of Ship Production*, 23(2), 89–93.

[58] Y.P. Yang and P. Dong (2012). Buckling distortion and mitigation techniques for thin-section structures. *Journal of Materials Engineering and Performance*, 21(2), 153–160.

[59] Y. Tajiama, S. Rashed, Y. Okumoto, Y. Katayama and H. Murakawa (2007). Prediction of welding distortion and panel buckling of car carrier decks using database generated by FEA. *Transactions of JWRI*, 36(1), 65–71.

[60] D. Deng and H. Murakawa (2008). FEM prediction of buckling distortion induced by welding in thin plate panel structures. *Computational Materials Science*, 43, 591–607.

[61] D. Deng, H. Murakawa and W. Liang (2007). Numerical simulation of welding distortion in large structures. *Computational Methods Application Math*, 196(45), 4613–4627.

[62] J. Wang, X. Shi, H. Zhou, Z. Yang and J. Liu (2020). Dimensional precision controlling on out-of-plane welding distortion of major structures in fabrication of ultra large container ship with 20000TEU. *Ocean Engineering*, 199, 106993.

[63] H. Zhao, Q. Zhang, Y. Niu, S. Du, J. Lu, H. Zhang and J. Wang (2020). Influence of triangle reinforcement plate stiffeners on welding distortion mitigation of fillet welded structure for lightweight fabrication. *Ocean Engineering*, 213, 107650.

[64] H. Zhou, J. Wang, H. Zhang, J. Liu and Z. Mo (2021). Prediction and mitigation of out-of-plane welding distortion of a typical block in fabrication of a semi-submersible lifting and disassembly platform. *Marine Structures*, 77, 102964.

[65] J. Wang, X. Yin and H. Murakawa (2013). Experimental and computational analysis of residual buckling distortion of bead-on-plate welded joint. *Journal of Materials Process Technology*, 213(8), 1447–1458.

[66] J. Wang, M. Shibahara, X. Zhang and H. Murakawa (2012). Investigation on twisting distortion of thin plate stiffened structure under welding. *Journal of Materials Process Technology*, 212(8), 1705–1715.

[67] J. Wang, N. Ma and H. Murakawa (2015). An efficient FE computation for predicting welding induced buckling in production of ship panel structure. *Mar Structure*, 41, 20–52.

[68] J. Wang, B. Yi and H. Zhou (2018). Framework of computational approach based on inherent deformation for welding buckling investigation during fabrication of lightweight ship panel. *Ocean Engineering*, 157, 202–210.

[69] J. Wang, S. Rashed and H. Murakawa (2014). Mechanism investigation of welding induced buckling using inherent deformation method. *Thin-Walled Structure*, 80, 103–19.

[70] J. Wang and B. Yi (2019). Benchmark investigation of welding-induced buckling and its critical condition during thin plate butt welding. *Journal of Manufacturing Science and Engineering*, 141(7), 071010.

[71] J. Wang, S. Rashed, H. Murakawa and Y. Luo (2013). Numerical prediction and mitigation of out-of-plane welding distortion in ship panel structure by elastic FE analysis. *Marine Structures*, 34, 135–155.

[72] W. Liang and D. Deng (2018). Influences of heat input, welding sequence and external restraint on twisting distortion in an asymmetrical curved stiffened panel. *Advanced Engineering Software*, 115, 439–451.

[73] J. Wang, X. Zhou and B. Yi (2022). Buckling distortion investigation during thin plates butt welding with considering clamping influence. *CIRP Journal of Manufacturing Science and Technology*, 37, 278–290

[74] S.M. Kelly, S.W. Brown, J.F. Tressler, R.P. Martukanitz and M.J. Ludwig (2009). Using hybrid laser arc welding to reduce distortion in ship panels. *Welding Journal*, 88(3), 32–36.

[75] J. Wang, B. Yi and H. Zhou (2018). Framework of computational approach based on inherent deformation for welding buckling investigation during fabrication of lightweight ship panel. *Ocean Engineering*, 157, 202–210.

[76] Y.P. Yang and P. Dong (2011). Buckling distortions and mitigation techniques for thin-section structures. *Journal of Materials Engineering and Performance*, 21, 153–160.

[77] D. Lee and S. Shin (2011). A study on control of buckling distortion at the aluminum GMA butt weldment by trail rolling method. *Proceedings of International Offshore Polar Engineering Conference*, Hawaii, Maui 105–110.

[78] D. Xu, X.S. Liu, P. Wang, J.G. Yang and H.Y. Fang (2009). New technique to control welding buckling distortion and residual stress with noncontact electromagnetic impact. *Science and Technology of Welding and Joining*, 14(8), 753–759.

[79] N. Ma, J. Wang and Y. Okumoto (2016). Out-of-plane welding distortion prediction and mitigation in stiffened welded structures. *International Journal of Advanced Manufacturing Technology*, 84, 1371–1389.

[80] J. Wang, S. Rashed, H. Murakawa and Y. Luo (2013). Numerical prediction and mitigation of out-of-plane welding distortion in ship panel structure by elastic FE analysis. *Marine Structures*, 34, 135–155.

[81] P. Michaleris and X. Sun (1997). Finite element analysis of thermal tensioning techniques mitigating weld buckling distortion. *Welding Journal*, 76(11), 451s–457s.

[82] M.V. Deo and P. Michaleris (2003). Mitigation of welding induced buckling distortion using transient thermal tensioning. *Science and Technology of Welding and Joining*, 8(1), 49–54.

[83] Y. Yang, R. Dull, C. Conrardy, N. Porter, P. Dong and T.D. Huang (2008). Transient thermal tensioning and numerical modeling of thin steel ship panel structures. *Journal of Ship Production*, 24(1), 37–49.

[84] Y.I. Burak, L.P. Besedina, Y.P. Romanchuk, A.A. Kazimirov, V.P. Morgun (1977). Controlling the longitudinal plastic shrinkage of metal during welding. *Avtomaticheskaya Svarka (Automatic Welding)*, 30(3), 27–29.

[85] Q. Guan, D.L. Guo and C.Q. Li (1994). Low stress non-distortion (LSND) welding – a new technique for thin materials. *Welding in the World*, 11(4), 231–237.

[86] Q. Guan (1999). A survey of development in welding stress and distortion controlling in aerospace manufacturing engineering in China. *Welding in the World*, 43(1), 64–74.

[87] P. Michaleris and X. Sun (1997). Finite element analysis of thermal tensioning techniques mitigating weld buckling distortion. *Welding Journal*, 76(11), 451.

[88] M.V. Deo and P. Michaleris (2003). Mitigation of welding induced buckling distortion using transient thermal tensioning. *Science and Technology of Welding and Joining*, 8(1), 49–54.

[89] Y.P. Yang and P. Dong (2011). Buckling distortions and mitigation techniques for thin-section structures. *Journal of Materials Engineering and Performance*, 21, 153–160.

[90] T.D. Huang, R. Dull, C. Conrardy, N. Porter, L. Decan, N. Evans, L. Kvidahl and P. Keene (2008). Transient thermal tensioning and prototype system testing of thin steel ship panel structures. *Journal of Ship Production*, 24(1), 25–36.

[91] J. Souto, E. Ares and P. Alegre (2015). Procedure in reduction of distortion in welding process by high temperature thermal transient tensioning. *Procedia Engineering*, 132, 732–739.

[92] W. Zhang, H. Fu, J. Fan, R. Li, H. Xu, F. Liu and B. Qi (2018). Influence of multi-beam preheating temperature and stress on the buckling distortion in electron beam welding. *Materials & Design*, 139, 439–446.

[93] M. Li, S. Ji, D. Yan and Z. Yang (2019). Controlling welding residual stress and distortion by a hybrid technology of transient thermal tensioning and trailing intensive cooling. *Science and Technology of Welding Joining*, 24, 527.

[94] B. Yi and J. Wang (2021). Mechanism clarification of mitigating welding induced buckling by transient thermal tensioning based on inherent strain theory. *Journal of Manufacturing Processes*, 68, 1280–1294.

[95] H. Zhou, B. Yi, C. Shen and J. Wang (2022). Mitigation of welding induced buckling with transient thermal tension and its application for accurate fabrication of offshore cabin structure. *Marine Structures*, 81, 103104.

[96] N. Ma, H. Huang and H. Murakawa (2015). Effect of jig constraint position and pitch on welding deformation. *Journal of Materials Processing Technology*, 221, 154–162.

[97] J. Wang, S. Rashed, H. Murakawa and Y. Luo (2013). Numerical prediction and mitigation of out-of-plane welding distortion in ship panel structure by elastic FE analysis. *Marine Structures*, 34, 135–155.

[98] A.M.A. Pazooki, M.J.M. Hermans and I.M. Richardson (2017). Finite element simulation and experimental investigation of thermal tensioning during welding of DP600 steel. *Science and Technology Welding and Joining*, 22, 7–21.

[99] G.S. Schajer and C.O. Ruud (2013). Overview of residual stresses and their measurement, in practical residual stress measurement methods. *Wiley*, Hoboken.

[100] W.C. Woo, G.B. An, E.J. Kingston, et al. (2013). Through-thickness distributions of residual stresses in two extreme heat-input thick welds: a neutron diffraction, contour method and deep hole drilling study. *Acta Material*, 61(10), 3564–3574.

[101] W.C. Woo, G.B. An, C.E. Truman, W. Jiang and M.R. Hill (2016). Two-dimensional mapping of residual stresses in a thick dissimilar weld using

contour method, deep hole drilling, and neutron diffraction. *Journal of Materials Science*, 51, 10620–10631.

[102] W.C. Woo, V.T. Em, P. Mikula, G.B. An and B.S. Seong (2011). Neutron diffraction measurements of residual stress in a 50 mm thick weld. *Materials Science and Engineering A*, 528(12), 4120–4124.

[103] J. Balakrishnan, A.N. Vasileiou, J.A. Francis, et al. (2018) Residual stress distributions in arc, laser and electron-beam welds in 30 mm thick SA508 steel: a cross-process comparison. *International Journal of Pressure Vessels and Piping*, 162, 59–70.

[104] A.N. Vasileiou, M.C. Smith, J.A. Francis, D.W. Rathod, J. Balakrishnan and N.M. Irvine (2019). Residual stresses in arc and electron-beam welds in 130 mm thick SA508 steel: part 2-measurements. *International Journal of Pressure Vessels and Piping*, 172, 379–390.

[105] M.E. Kartal, Y.H. Kang, A.M. Korsunsky, A.C.F. Cocks and J.P. Bouchard (2016). The influence of welding procedure and plate geometry on residual stresses in thick components. *International Journal of Solids and Structures*, 80, 420–429.

[106] J.U. Park, G.B. An, W.C. Woo, J.H. Choi and N.S. Ma (2014). Residual stress measurement in an extra thick multi-pass weld using initial stress integrated inherent strain method. *Marine Structures*, 39, 424–437.

[107] B. Qiang, Y. Li, C. Yao, X. Wang and Y. Gu (2018). Through-thickness distribution of residual stress in Q345 butt-joint welded steel plates. *Journal of Materials Processing Technology*, 251, 54–64.

[108] W. Jiang, Y. Wan, S. Tu, et al. (2022). Determination of the through-thickness residual stress in thick duplex stainless steel welded plate by wavelength-dependent neutron diffraction method. *International Journal of Pressure Vessels and Piping*, 196, 104603.

[109] D.J. Smith, G. Zheng, P.R. Hurrell, et al. (2014). Measured and predicted residual stresses in thick section electron beam welded steels. *International Journal of Pressure Vessels and Piping*, 120–121(1), 66–79.

[110] S.Y. Hwang, Y. Kim, and J.H. Lee. Finite element analysis of residual stress distribution in a thick plate joined using two-pole tandem electro-gas welding. *Journal of Materials Processing Technology*, 2016, 229: 349–360.

[111] Y. Wan, W. Jiang, J. Li, et al. (2017). Weld residual stress in a thick plate considering back chipping: neutron diffraction, contour method and finite element simulation study. *Materials Science and Engineering A*, 699, 62–70.

[112] W. Jiang, W.C. Woo, Y. Wan, Y. Luo, X. Xie, S. Tu (2017). Evaluation of through-thickness residual stress by Neutron diffraction and finite element method in thick welded plates. *Journal of Pressure Vessel Technology*, 139(3), 1–10.

[113] H. Zhou, Q. Zhang, B. Yi and J. Wang (2020). Hardness prediction based on microstructure evolution and residual stress evaluation during high tensile thick plate butt welding. *International Journal of Naval Architecture and Ocean Engineering*, 12, 146–156.

[114] H. Murakawa, N.S. Ma and H. Huang (2015). Iterative substructure method employing concept of inherent strain for large-scale welding problems. *Welding in the World*, 59, 53–63.

[115] K. Ikushima and M. Shibahara (2014). Prediction of residual stresses in multi-pass welded joint using idealized explicit FEM accelerated by a GPU. *Computational Materials Science*, 93, 62–67.

[116] X. Pu, C. Zhang, S. Li and D. Deng (2017). Simulating welding residual stress and deformation in a multi-pass butt-welded joint considering balance between computing time and prediction accuracy. *The International Journal of Advanced Manufacturing Technology*, 93, 2215–2226.

[117] S. Ren, S. Li, Y. Wang, D. Deng and N.S. Ma (2019). Predicting welding residual stress of a multi-pass P92 steel butt-welded joint with consideration of phase transformation and tempering effect. *Journal of Materials Engineering and Performance*, 28(12), 7452–7463.

[118] N.S. Ma, Z. Cai, H. Huang, D. Deng, H. Murakawa and J. Pan (2015). Investigation of welding residual stress in flash-butt joint of U71Mn rail steel by numerical simulation and experiment. *Materials and Design*, 88, 1296–1309.

[119] W. Jiang, W. Chen, W.C. Woo, S.T. Tu, X.C. Zhang and V. Em (2018). Effects of low-temperature transformation and transformation-induced plasticity on weld residual stresses: numerical study and neutron diffraction measurement. *Materials and Design*, 147, 65–79.

[120] J.U. Park, G.B. An and W.C. Woo (2018). The effect of initial stress induced during the steel manufacturing process on the welding residual stress in multi-pass butt welding. *International Journal of Naval Architecture and Ocean Engineering*, 10(2), 129–140.

[121] M. Xu, H. Lu, C. Yu, J. Xu and J. Chen (2013). Finite element simulation of butt welded 2·25Cr–1·6W steel pipe incorporating Bainite phase transformation. *Science and Technology of Welding & Joining*, 18, 184–190.

[122] A.L. Gurson (1977). Continuum theory of ductile rupture by void nucleation and growth: part 1-yield criteria and flow rules for porous ductile media. *Journal of Engineering Materials and Technology*, 99(1), 297–300.

[123] V. Tvergaard (1981). Influence of voids on shear band instabilities under plane strain conditions. *International Journal of Fracture*, 17(4), 389–407.

[124] V. Tvergaard (1982). On localization in ductile materials containing spherical voids. *International Journal of Fracture*, 18(4), 237–252.

[125] C. Chu and A. Needleman (1980). Void nucleation effects in biaxial stretched sheets. *Journal of Engineering Materials and Technology*, 102(3), 249–256.

[126] M. Achouri, G. Germain, P.D. Santo and D. Saidane (2013). Numerical integration of an advanced Gurson model for shear loading: application to the blanking process. *Computational Materials Science*, 72, 62–67.

[127] M. Rakin, B. Medjo, N. Gubeljak and A. Sedmak (2013). Micromechanical assessment of mismatch effects on fracture of high-strength low alloyed steel welded joints. *Engineering Fracture Mechanics*, 109, 221–235.

[128] R. Chhibber, P. Biswas, N. Arora, S.R. Gupta and B.K. Dutta (2011). Micromechanical modelling of weldments using GTN model. *International Journal of Fracture*, 167, 71–82.

[129] F.G. Carlos, F.E.I. Saavedra and H.A. Marie (2018). Damage characterization in a ferrite steel sheet: experimental tests, parameter identification and numerical modeling. *International Journal of Solids and Structures*, 155, 109–122.

[130] M. Marteleur, J. Leclerc, M.S. Colla, V.D. Nguyen and T. Pardoen (2021). Ductile fracture of high strength steels with morphological anisotropy, Part I: characterization, testing, and void nucleation law. *Engineering Fracture Mechanics*, 244, 107569.

[131] B. Qiang and X. Wang (2019). Ductile crack growth behaviors at different locations of a weld joint for an X80 pipeline steel: a numerical investigation using GTN models. *Engineering Fracture Mechanics*, 213, 264–279.

[132] X. Wu, J. Shuai, K. Xu, Z. Lv and K. Shan (2020). Determination of local true stress-strain response of X80 and Q235 girth-welded joints based on digital image correlation and numerical simulation. *International Journal of Pressure Vessels and Piping*, 188, 104232.

[133] R. Gadallah, N. Osawa and S. Tanaka (2017). Evaluation of stress intensity factor for a surface cracked butt welded joint based on real welding residual stress. *Ocean Engineering*, 138, 123–139.

[134] J.H. Lee, B.S. Jang, H.J. Kim, S.H. Shim and S.W. Im (2020). The effect of weld residual stress on fracture toughness at the intersection of two welding lines of offshore tubular structure. *Marine Structures*, 71, 102708.

[135] J. Pan, S. Chen, Z. Lai, Z. Wang, J. Wang and H. Xie (2017). Analysis and fracture behavior of welded box beam-to-column connections considering residual stresses. *Construction and Building Materials*, 154, 557–566.

[136] Z. Xiaomin, X.U. Jie, L.I. Pengpeng, et al. (2017). Effect of residual stresses on ductile fracture of pipeline steels[J]. *Hanjie Xuebao/Transactions of the China Welding Institution*, 38(5), 44–48.

[137] X. Ren, Z. Zhang and B. Nyhus (2010). Effect of residual stresses on ductile crack growth resistance. *Engineering Fracture Mechanics*, 77, 1325–1337.

[138] V. Rudnev, D. Loveless, R.L. Cook, et al. (2002). Handbook of induction heating, Second Edition [M], CRC Press, London.

[139] A. Turk, G.R. Joshi, M. Gintalas, M. Callisti and E.I. Galindo-Nava (2020). Quantification of hydrogen trapping in multiphase steels: part II – Effect of austenite morphology. *Acta Materialia*, 197, 253–268.

[140] S. Wu, C. Zhang, L. Zhu, Q. Zhang and X. Ma (2020). In-depth analysis of intragranular acicular Ferrite three-dimensional morphology. *Scripta Materialia*, 185, 61–65.

[141] Y. Wang, Y. Adachi, K. Nakajima and Y. Sugimoto. Quantitative three-dimensional characterization of Pearlite spheroidization. *Acta Materialia*, 2010, 58(14), 4849–4858.

[142] E. Keehan, L. Karlsson, H.K.D.H. Bhadeshia and M. Thuvander (2008). Three-dimensional analysis of coalesced Bainite using focused ion beam tomography. *Materials Characterization*, 59(7), 877–882.

[143] S. Morito, Y. Edamatsu, K. Ichinotani, et al. (2013). Quantitative analysis of three-dimensional morphology of Martensite packets and blocks in iron-carbon-manganese steels. *Journal of Alloy and Compounds*, 577, 587–592.

[144] J.C.F. Jorge, L.F.G. Souza, M.C. Mendes, et al. (2021). Microstructure characterization and its relationship with impact toughness of C–Mn and high strength low alloy steel weld metals–a review. *Journal of Materials Research and Technology*, 10, 471–501.

[145] J. Liao, K. Ikeuchi and F. Matsuda (1998). Toughness investigation on simulated weld HAZs of SQV-2A pressure vessel steel. *Nuclear Engineering and Design*, 183, 9–20.

[146] Y. Chen, C. Yang, H. Chen, H. Zhang and S. Chen (2015). Microstructure and mechanical properties of HSLA thick plates welded by novel double-sided gas metal arc welding. *International Journal of Advanced Manufacturing Technology*, 78, 457–464.

Printed in the United States
by Baker & Taylor Publisher Services